T0133804

Current Concepts in General Surgery: A Resident Review

William R. Wrightson, M.D.
Cardiothoracic Surgery Fellow
University of Louisville
Louisville, Kentucky, U.S.A.

LANDES
BIOSCIENCE
GEORGETOWN, TEXAS
U.S.A.

VADEMECUM
Current Concepts in General Surgery: A Resident Review
LANDES BIOSCIENCE
Georgetown, Texas U.S.A.

Cover Artwork: Kristen Shumaker

Please address all inquiries to the Publisher:
Landes Bioscience, 810 S. Church Street, Georgetown, Texas, 78626 U.S.A.
Phone: 512/ 863 7762; FAX: 512/ 863 0081

ISBN: 1-57059-693-X

Library of Congress Cataloging-in-Publication Data

Current concepts in general surgery : a resident review / [edited by] William R. Wrightson.
    p. ; cm. -- (Vademecum)
 Includes bibliographical references.
 ISBN 1-57059-693-X
 1. Surgery, Operative--Handbooks, manuals, etc. I. Wrightson, William R. II. Title. III. Series.
 [DNLM: 1. Surgical Procedures, Operative--Handbooks. 2. Clinical Medicine--methods--Handbooks. 3. Diagnostic Techniques and Procedures--Handbooks. 4. Signs and Symptoms--Handbooks. WO 39 C9756 2006]
RD32.C87 2006
617'.91--dc22

                              2006003718

Dedication

This book is dedicated to my Father. Although you couldn't be there with me through my surgical training or walk down the path I have chosen, you were always my inspiration and my guide. My faith, hope, dedication, and love stem from your tutelage and belief in striving for excellence.

Contents

19. Cardiothoracic Surgery 167

William R. Wrightson

Editor

William R. Wrightson, M.D.
Cardiothoracic Surgery Fellow
University of Louisville
Louisville, Kentucky, U.S.A.
Chapters 1, 3, 7, 8, 10, 12, 13, 14, 15, 18, 19, 20

Contributors

Timothy K. Bullock, M.D.
General Surgery Private Practice
Birmingham, Alabama, U.S.A.
Chapter 16

Phillip T. Burch, M.D.
General Surgery Resident
University of Louisville
Louisville, Kentucky, U.S.A.
Chapter 2

Mary B. Carter, M.D., Ph.D.
Assistant Professor
University of Louisville
Louisville, Kentucky, U.S.A.
Chapter 1

Steven R. Casos, M.D.
General Surgery Resident
University of Louisville
Louisville, Kentucky, U.S.A.
Chapter 12

C. Adam Conn, M.D.
General Surgery Resident
University of Louisville
Louisville, Kentucky, U.S.A.
Chapter 20

Brian R. Davis, M.D.
General Surgery Resident
University of Louisville
Louisville, Kentucky, U.S.A.
Chapter 4

James C. Dodds, M.D.
Orthopedic Surgery Chief Resident
University of Louisville
Louisville, Kentucky, U.S.A.
Chapter 21

Stephen M. Girard, M.D.
General Surgery Private Practice
Pueblo, Colorado, U.S.A.
Chapter 11

Steven L. Goudy, M.D.
Otolaryngologic Surgery Chief Resident
University of Louisville
Louisville, Kentucky, U.S.A.
Chapter 22

Monica S. Hall, M.D.
General Surgery Resident
University of Louisville
Louisville, Kentucky, U.S.A.
Chapter 23

Robert C. Kanard, M.D.
General Surgery Resident
University of Louisville
Louisville, Kentucky, U.S.A.
Chapter 5

Vincent C. Lusco III, M.D.
General Surgery Chief Resident
University of Louisville
Louisville, Kentucky, U.S.A.
Chapter 17

Robert I. Oliver, Jr., M.D.
General Surgery Chief Resident
University of Louisville
Louisville, Kentucky, U.S.A.
Chapter 6

S. Matthew Rose, M.D.
Orthopedic Surgery Resident
University of Louisville
Luisville, Kentucky, U.S.A.
Chapter 21

Alina D. Sholar, M.D.
General Surgery Resident
University of Louisville
Louisville, Kentucky, U.S.A.
Chapter 14

Sara Elizabeth Snell, M.D.
General Surgery Chief Resident
University of Louisville
Louisville, Kentucky, U.S.A.
Chapter 9

Dan N. Tran, M.D.
General Surgery Resident
University of Louisville
Louisville, Kentucky, U.S.A.
Chapter 13

CHAPTER 1

Basic Physiology

Mary B. Carter and William R. Wrightson

Fluids and Electrolytes

Mary B. Carter

Total Body Water can be described as a percentage of total body weight (TBW). Based on various other factors including age, fat composition and sex (Table 1.1).

The fluid is divided into compartments including intracellular (66% TBW) and extracellular (33% TBW). The extracellular fraction is then divided into intravascular (7% of TBW) and extravascular (15% TBW).

Water Balance

Individuals take in 2000 to 2500 mL of fluid each day. Of the total, 1500 mL in the form of oral liquids and the rest is from solid food or products of oxidation. There is 800 to 1500 mL of urine, 250 mL of stool and 600-900 mL of insensible losses per day to maintain this balance. Normal osmotic pressure 290-310 mOsm.

Extracellular Volume Loss

The causes of extracellular volume loss include vomiting, NGT suction, diarrhea, fistula drainage, third spacing due to infection or burns. The signs of this fluid loss are oliguria, sleepiness, coma, tachycardia, hypotension and decreased body temperature. Management includes IV fluid resuscitation with normal saline mixed with KCl if due to GI losses, lactated ringers if due to burn.

Electrolyte Abnormalities

Sodium

Hypernatremia (>150 mEq/L) results from a total sodium excess or relative fluid loss. This can be from prolonged administration of excess sodium or excessive loss of free water. Hyperglycemia can also act as an osmotic diuretic increasing free water loss. Drugs such as amphoteracin B, methicillin and gentamicin can cause a nephrogenic diabetes and subsequent free water loss.

Patients present with restlessness, delirium, weakness, tachycardia, decreased saliva, dry sticky mucous membranes, flushed skin, oliguria, fever and or heatstroke. Treatment includes administration of free water and minimizing sodium intake (<3 g/day). One half of the free water deficit should be administered over a 24 hour period to minimize edema.

Hyponatremia is a common finding in the surgical patient and is usually the result of excessive administration of H_2O or hypotonic fluids. Patients may be symptomatic with levels <130 mEq/L and develop seizures <120 mEq/L. The presentation is with muscle twitching, increased DTRs, increased ICP, convulsions, hypertension, increased salivation, watery diarrhea, oliguria. In many cases there are

Current Concepts in General Surgery: A Resident Review, edited by William R. Wrightson. ©2006 Landes Bioscience.

Table 1.1. Percent total body weight as water

	Percent Body Water
Infants	75-80%
Children	65%
Lean adult males	60%
Lean adult females	50%
Obese adults	40-45%
Elderly adult	47-52%

no symptoms until sodium below 120 mEq/L. Symptoms appear earlier in head injured patients causing increased ICP. In severe hyponatremia, small amounts of hypertonic NaCl are administered IV, followed by normal saline IV. (Note: serum sodium should not increase more than 12 mEq/L per liter in the first 24 hours.) If volume is expanded, restrict free water intake to 1500 mL per 24 hours. Cirrhotic patients may also require spironolactone or furosemide to mobilize free water.

Potassium

Hyperkalemia results from renal failure, acidosis, iatrogenic causes. Signs include nausea, vomiting, diarrhea, peaked T waves, widened QRS complex, depressed S-T segments. Fatal ventricular arrhythmias can result demanding rapid management. Diuresis with infusion of NaCl infusion with furosemide 40-80 mg can also decrease potassium levels over time. Hemodialysis may be required in extreme cases (Table 1.2).

Hypokalemia results from loop diuretics, vomiting, alkalosis, iatrogenic causes. It can also lead to atrial and ventricular arrhythmias but usually not until levels fall below 3 mEq/L. Treatment is with KCl IV if adequate urinary output is established (not to exceed 20 mEq per hour unless the patient is on a cardiac monitor). Each 10 mEq of KCl will only raise the serum potassium level by 0.1 mEq/L.

Table 1.2. Treatment of hyperkalemia

Treatment	Effect
Calcium gluconate (10%)	Reduces electrical excitability of the heart
Bicarbonate	Drives potassium into cells by inducing transient alkalosis
Insulin/D50	Drives potassium into cells
Sodium polystyrene sulfonate (Kayexlate)	Binds potassium (0.1 mEq per gram)

Calcium

Hypocalcemia is caused by necrotizing fasciitis, renal failure, intestinal fistulas, post-op from excision of parathyroid adenoma, hypoparathyroidism and magnesium depletion. In acute pancreatitis calcium is sequestered in saponified fat. Hypocalciemia can occur after massive transfusions and should routinely be replaced if >6 units of blood are given. Signs include hyperactive DTRs, muscle spasms, positive Chvostek sign, tetany, carpopedal spasm, convulsions, prolongation of the Q-T interval, numbness and tingling of the circumoral region and the tips of the digits. Treatment is by managing the underlying cause, administration of IV calcium gluconate or calcium chloride.

Hypercalcemia is found in hyperparathyroidism and malignancies with bony metastases and paraneoplastic syndromes. Signs include easy fatigue, weakness, anorexia, nausea, vomiting, weight loss, stupor, coma, body aches, headaches, thirst, polydipsia and polyuria. Treatment is with normal saline fluid resuscitation, furosemide, sodium phosphate, corticosteroids, plicamycin, calcitonin. Surgical excision is employed for patients with hyperparathyroidism in hypercalcemic crisis.

Magnesium

Only 1% of magnesium is found in the extracellular space with the remaining in bone and muscle. Hypomagnesemia can result in hypermetabolic states, muscle wasting, starvation, malabsorption, fistula, primary aldosteronism, chronic alcoholism. Signs include hyperactive tendon reflexes, muscle tremors, tetany, positive Chvostek sign, delirium and convulsions. Treatment is with intravenous or oral administration of magnesium.

Hypermagnesemia is fairly rare with causes including renal failure, burns, massive trauma, volume deficit and acidosis. Patients will have lethargy, weakness, loss of DTRs, coma, paralysis, respiratory or cardiac arrest, widened QRS complex, elevated T waves and/or increased P-R interval. Treatment is with a brisk diuresis, hydration and correction of underlying disease process.

Hormones

Mary B. Carter

Hormones Secreted from the Anterior Pituitary

Adrenocorticotropic Hormone (ACTH)

ACTH is a 39 amino acid hormone that simulates cortisol production by the adrenal gland. It also stimulates secretion and growth of both the zona fasciculate and zona reticularis of the adrenal cortex and provides negative feedback to hypothalamic CRH release.

Thyroid Stimulating Hormone (TSH)

TSH is a glycoprotein with 211 amino acids and stimulates production and release of thyroxine (T_4) and 3,5,3'-triiodothyronine (T_3). TSH stimulates iodide transport into thyroid cells and growth of the thyroid gland itself and provides negative feedback to TRH release from the hypothalamus.

1

Growth Hormone (GH)

GH is an amino acid peptide hormone that stimulates longitudinal growth of bone and insulin secretion by the pancreas. It antagonizes insulin effects on sugar uptake and fatty acid release but enhances the anabolic effect of insulin on amino acid uptake. Growth hormone produces a positive nitrogen and phosphorus balance. It stimulates hepatocyte growth and adipocyte metabolism with mobilization of free fatty acids from adipose tissue, favoring ketogenesis. GH increases hepatic glucose output, promotes hyperglycemia.

It has significant anabolic effects in burn patients and induces release of somatomedins, or insulin-like growth factors IGF-1 and IGF-2, to regulate metabolism and produce anabolism. Somatomedins produce feedback inhibition of GH release from the pituitary and stimulate somatostatin release from the hypothalamus

Luteinizing Hormone (LH)

It stimulates Leydig cells in males to produce testosterone. Midcycle surge in females causes follicular rupture, ovulation and establishment and maintenance of the corpus luteum.

Follicle-Stimulating Hormone (FSH)

FSH promotes spermatogenesis in Sertoli cells in males and stimulates maturation of the Graafian follicle and its production of estradiol in females.

Prolactin

Prolactin is synthesized and released in response to sucking of the nipple. Secretion is also increased by exercise or surgical and psychological stress. Acts in the breast to initiate and sustain lactation and inhibits the effects of gonadotropins FSH and LH.

Hormones Secreted by the Posterior Pituitary Gland

Antidiuretic Hormone (ADH, Vasopressin)

ADH is a 9 amino acid hormone synthesized and released in response to rise in plasma osmolality above 285 mOsm or a decrease in circulating blood volume by 5% or more. It stimulates sodium and chloride reabsorption in the thick ascending loop of Henle and enhances permeability to water in the collecting ducts of the renal medulla, decreasing water loss in urine

Oxytocin

Oxytocin is a 9 amino acid hormone secreted in response to distention of the vagina or uterus or by sucking of the nipples. It stimulates uterine contraction during labor and milk ejection by myoepithelial cells of the mammary ducts during lactation.

Peripherally Secreted Hormones

Thyroxine (T_4) and Triiodothyronine (T_3)

Synthesized in the thyroid gland. They stimulate oxygen consumption, increase body heat, metabolism and heart rate. There is an increase in the dissociation of oxygen from hemoglobin by increasing red cell 2,3-diphosphoglycerate (2,3-DPG)

and an increase in the activity of the Na$^+$-K$^+$ ATPase in many tissues thus increasing energy consumption.

Mitochondrial protein synthesis and carbohydrate absorption from the GI tract are enhanced.

It increases the number and affinity of beta-adrenergic receptors in the heart, thus increasing the cardiac sensitivity to catacholamines. It also potentiates the effect of GH on tissue.

Parathyroid Hormone (PTH)

Synthesized and released from chief cells in the parathyroid gland. PTH promotes increase in serum calcium by increased release from bones, decreased renal excretion, and enhanced absorption from the intestine via vitamin D$_3$. It stimulates osteoclasts and inhibits osteoblasts, thus increasing resorption of calcium and phosphate from bone and inhibits phosphate reabsorption in the renal tubule, thus increasing phosphate excretion in the urine. PTH stimulates hydroxylation of 25-hydroxyvitamin D to 1,25-dihydroxyvitamin D in the kidney and renal secretion of bicarbonate.

Calcitonin

Calcitonin is secreted from the parafollicular cells or clear cells (ultimobranchial bodies) in the thyroid gland and inhibits bone resorption and produces hypocalcemia in experimental animals. Its secretion results in decreases in serum phosphate levels but it has not been demonstrated to be important in the basal control of serum calcium during normal homeostasis in humans. Salmon calcitonin is more than 20 times more active in humans than human calcitonin.

Cortisol and Other Glucocorticoid Hormones

Produced and secreted from the adrenal cortex and stimulates release of glucagons. It increases blood glucose by increasing gluconeogenesis in the liver and decreasing insulin-stimulated glucose uptake in peripheral tissues and decreases peripheral amino acid uptake and protein synthesis but increases hepatic uptake of amino acids. It also increases protein catabolism and peripheral lypolysis.

Immunologic effects include a decrease in interleukin-2 production, inhibition of B-cell activation and proliferation, inhibition of monocyte and neutrophil migration to areas of inflammation and decreased histamine release and histamine-induced lysosomal degranulation by mast cells.

Stimulates angiotensinogen release and decreases the release of the vasodilator prostaglandin I$_2$. Cortisol inhibits collagen formation, fibroblast activity, and formation of bone by osteoblasts with a decrease plasma calcium levels by inhibiting osteoclast formation and activity. It provides feedback inhibition of ACTH and CRH release in the anterior pituitary and hypothalamus, respectively

Aldosterone

Aldosterone is secreted from the zona glomerulosa of the adrenal gland in response to hyperkalemia, circulating angiotensin II, surgery, physical trauma, hemorrhage, or anxiety. It stimulates sodium retention and potassium and hydrogen ion secretion in the distal convoluted tubule of the kidney and promotes sodium absorption by the intestinal mucosa, sweat glands, and salivary glands.

1

Dehydroepiandrosterone (DHEA)

DHEA is the major C-19 sex steroid produced by the zona reticularis of the adrenal cortex that acts as a weak androgen which promotes virilization in women. During fetal life, contributes to development of the male external genitalia, vas deferens, epididymis, seminal vesicles, and prostate.

Insulin

Insulin is a polypeptide hormone with two chains of amino acids linked by disulfide bridges synthesized in and secreted by the β cells of the islets of Langerhans in response to glucose. Insulin facilitates the entry of glucose into cells by increasing the number of glucose transporters in the cell membranes, increases transport of amino acids and K^+ into cells and increases the activity of Na^+-K^+ ATPase in cell membranes. Insulin also results in stimulation of protein synthesis, lipogenesis and inhibits protein degradation.

Glucagon

Glucagon is a linear polypeptide hormone with 29 amino acids synthesized in the α cells of the islets of Langerhans and secreted into the portal vein in response to fasting, starvation, glucogenic amino acids, CCK, gastrin, cortisol, exercise, infections, stress, beta-adrenergic stimulators, theophylline, or acetylcholine. Secretion is inhibited by glucose, insulin, somatostatin, secretin, free fatty acids, ketones, phenytoin, alpha-adrenergic stimulators, GABA. It stimulates gluconeogenesis, glycogenolosis, lypolysis, and ketogenesis in the liver.

Glucagon requires glucocorticoids to exert its gluconeogenetic effects during fasting. Large doses of exogenous glucagon exert a positive inotropic effect on the heart.

Somatostatin

Polypeptide hormone synthesized in the Δ cells of the islets of Langerhans that inhibits the secretion of insulin and glucagons. It causes hyperglycemia, slows gastric emptying, decreases gastric acid secretion. Somatostatin inhibits CCK secretion leading to gallstones in patients with somatostatin secreting pancreatic tumors.

Testosterone

Testosterone is secreted by Leydig cells in males, converted to more active form, dihydrotestosterone. It stimulates growth of the penis and scrotum and promotes development of facial, axillary, and pubic hair. It also inhibits GnRH release from the hypothalamus.

Estrogen/Estradiol

Estrogen exerts negative feedback on release of LH and FSH by the pituitary except in mid-cycle when a surge of LH and FSH occur in response to estrogen. It increases the motility of fallopian tubes and stimulates endometrial growth during first half of menstrual cycle, growth and enlargement of breasts, and pigmentation of areolas, and enlargement of the uterus and vagina.

Progesterone

Progesterone exerts negative feedback on release of LH and FSH by the pituitary except in mid-cycle when progesterone surge enhances release of LH and FSH, leading to ovulation. It stimulates development of lobules and alveoli in the breast and causes progestational changes in the endometrium. Progesterone is responsible for cyclic changes in the cervix and vagina. It decreases the excitability of myometrial cells and their sensitivity to oxytocin.

References

1. Principles of Surgery. 7th ed. In: Schwartz SI, Shires GT, Spencer FC et al, eds. New York: McGraw-Hill Health Professions Division, 1999: Chapters 1 and 2.
2. Review of Medical Physiology. 20th ed. In: Ganong WF, ed. New York: Lange Medical Books/McGraw-Hill Medical Publishing Division, 2001: Chapters 18-24.
3. Textbook of Surgery: The Biological Basis of Modern Surgical Practice. 15th ed. In: Sabiston Jr DC, Lyerly HK, eds. Philadelphia: W. B. Saunders Company, 1997: Chapters 4, 6, 23-25.

Acid-Base Abnormalities

William R. Wrightson

Background

Acid-base balance is required to maintain a physiologic milieu compatible with life. The pH, or hydrogen ion concentration, is kept between 7.35 and 7.45 through an intricate system of buffers. The pH is typically expressed through the famous Henderson-Hasselbach equation:

The most significant buffer utilized is the bicarbonate system. Hydrogen ion concentration is dependent on the ratio of bicarbonate to carbon dioxide ($PaCO_2$) and not the absolute concentrations.

Acidosis and alkalosis refer to the presence of acids or bases and do not refer to alterations in the pH. On the other hand, acidemia and alkalemia refer to a deviation of the pH from normal values (Table 1.3).

Respiratory Acidosis

This is the result of a primary decrease in the respiratory rate and retention of CO_2 ($PaCO_2$ above 40). This results in an increase in hydrogen concentration and therefore a decrease in pH. Causes include COPD, narcotic overdose, flail chest, pneumothorax and severe obesity. Renal compensation results in increased excretion of H^+ and NH_4^+ and retention of HCO_3^-.

Respiratory Alkalosis

This results in an increase in the respiratory rate with a loss of CO_2 ($PaCO_2 < 40$) Causes are pain, emotional excitement, hypoxia, fever, respiratory compensation in response to metabolic acidosis. Renal compensation is decreased H^+ excretion and decreased HCO_3^- reabsorption.

Table 1.3. Acid-base disorders

Disorder	pH	H^+	CO_2	HCO_3^-
Metabolic Acidosis	Decreased	Increased	Decreased	Decreased
Metabolic Alkalosis	Increased	Decreased	Increased	Increased
Respiratory Acidosis	Decreased	Increased	Increased	Decreased
Respiratory Alkalosis	Increased	Decreased	Decreased	Increased

Metabolic Acidosis

This results from increased production of acid or increased loss of bicarbonate. When the pH falls below 7.1 ventricular arrhythmias, decreased contractility, and decreased peripheral vascular resistance may result. This may also result in increased intracranial pressure.

Metabolic Acidosis with Normal Anion Gap

Results from a loss of bicarbonate. This can be from diarrhea, small bowel fistula, ureterosigmoidostomy or renal tubular acidosis. Ureterosigmoidostomy causes hyperchloremic metabolic acidosis.

Metabolic Acidosis with Increased Anion Gap

Retention of acids results in a negative base excess. Causes include lactic acidosis in shock, diabetic ketoacidosis, renal failure, starvation, alcoholic ketoacidosis, ethylene glycol, salicylates and paraldehyde.

Metabolic Alkalosis

Metabolic alkalosis is the result of an increase in pH and bicarbonate concentration. This may be due to an excessive loss of hydrogen ions or increased administration of bicarbonate. This is most commonly the result of urinary and GI losses. Contraction alkalosis results from excessive diuresis with maintenance of bicarbonate concentration.

- **Metabolic alkalosis chloride responsive**
  Results from a loss of acid (vomiting, NGT suction, villous adenoma of the colon) or potassium depletion (diuretics)
- **Metabolic alkalosis chloride resistant**
  There is a maximal reabsorption of bicarbonate with excessive loss of chloride in the urine (primary hyperaldosteronism, Cushing's disease, exogenous corticosteroids, renal compensation to chronic respiratory acidosis)

CHAPTER 2

Surgical Immunology

Phillip T. Burch

Immunology

Introduction

Immunology is the study of the response of the body to substances it recognizes as foreign. The immune system can be separated into two divisions, innate and acquired immunity. The innate immune system responds quickly to foreign invaders; however it is incapable of amplifying its response to repeated exposures to that same antigen. On the other hand, the acquired or adaptive system as it is also known requires more time to respond; however it recognizes specific features of the antigen and upon subsequent exposures responds more efficiently. Innate immunity appears to have developed prior to the adaptive system. In addition to effector mechanisms that are characteristic of the adaptive immune response, it is capable of recruiting the innate system to amplify the acquired response. Due to its ability to change, the acquired response is the primary focus of the field of immunology.

Innate Immunity

Immune defenses that do not change their response to repeated exposure to an antigen comprise innate immunity. The timing and magnitude of the response is therefore the same regardless of the number of exposures to the antigen. The three classes of infectious organisms are bacteria, viruses, and parasites (protozoa, fungi, helminthes, and nematodes). Despite its limitations the innate system is equipped to respond to each type of pathogen.

The first lines of defense against microbial challenges are the mechanical barriers. The skin is a barrier impermeable to microbes that prevents them from entering the body. Secretions of apocrine and eccrine glands containing lactic acid and fatty acids are also on the surface of the skin and inhibit the growth of microorganisms. Similarly, the mucosal surfaces of the body serve as a barrier to infection and secrete substances (saliva, gastric acid, tears, etc.) that impede microbial growth. Additionally, actions like sneezing, coughing, increased peristalsis, and ciliary movement actively expel irritants and infectious agents.

Complement Cascade

If a foreign agent penetrates the mechanical barriers, it encounters the complement system. The complement activation cascade is initiated by two different mechanisms. The first method discovered is called the classical pathway although the alternative pathway appears to have developed first since the initiation of the classical pathway requires the participation of the acquired immune system. The alternative pathway can be considered innate immunity, and the classical pathway can be considered acquired immunity, however both will be discussed here.

Current Concepts in General Surgery: A Resident Review, edited by William R. Wrightson. ©2006 Landes Bioscience.

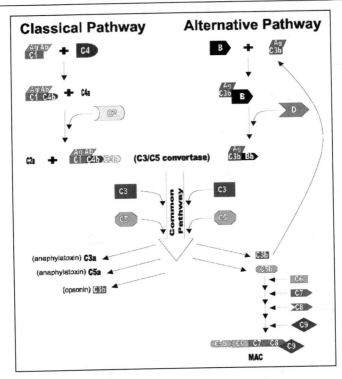

Figure 2.1. Complement cascade.

As Figure 2.1 illustrates, the classical pathway begins with C1 binding to immune complexes (i.e., a composite of antigen and antibodies). C1 is composed of C1q, C1r, and C1s. C1 must bind to two complement fixation sites for activation. These fixation sites are found on the Fc region of IgG or IgM. Thus activation requires two molecules of IgG because it is a monomer or one molecule of IgM since it is a pentamer. When C1q binds to the antibody, it is activated and converts C1r to an active state. Active C1r then activates C1s. Next active C1s cleaves C4 (note: products of complement cleavage are designated a and b, with a being the smaller fragment and b being the larger fragment) to C4a and C4b. C4a acts as a weak inflammatory mediator that increases local blood flow, vascular permeability, and capillary leak of fluid and protein. C4b covalently attaches to the surface of the antigen where it then binds to C2 making it susceptible to C1s. C1s then cleaves C2 producing C2b which remains bound to C4b forming the C4b,2b complex. This complex possesses the ability to process both C3 and C5 thus it is referred to as C3/C5 convertase. The C2b portion provides the enzymatic function while C4b keeps the complex anchored to the pathogen insuring the cleavage products are directed at the appropriate target. The C3/C5 convertase then cleaves multiple molecules of C3 producing large amounts of both C3a and C3b. C3a is a more active mediator of local inflammation than C4a. Production of large amounts of C3b is one of the

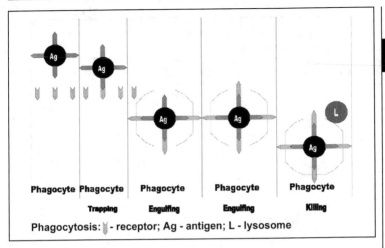

Figure 2.2. Process of phagocytotsis.

most important results of the activation of the complement cascade. C3b opsonizes the pathogen making it more readily phagocytized by macrophages and neutrophils (Fig. 2.2). C3b also binds to C5 making it susceptible to the action of the C3/C5 convertase producing C5a and C5b. C5a is the most potent of the mediators of local inflammation produced by the complement cascade. C5a and C3a, to a lesser degree, are chemotaxins for macrophages and neutrophils. The other product of C5 cleavage, C5b, binds to C6 forming C5b,6 which then binds to C7. The C7 portion of this complex has a hydrophobic site which inserts into the membrane of the pathogen anchoring C5b,6,7. Next C8 binds to C5b,6,7, and it also anchors the complex to the pathogen. Finally, the C8 portion of the C5b,6,7,8 complex binds to C9 and causes the polymerization of 10 to 16 C9 molecules into a porelike structure called the membrane attack complex (MAC). The MAC then disrupts the integrity of the pathogen's outer membrane allowing free passage of water and solutes across the membrane. This alters cellular homeostasis and eventually the cell lyses.

The activation of the alternative pathway requires no assistance. Spontaneous cleavage of C3 occurs at low levels in serum producing C3b. If an antigen is present, the C3b binds to its surface and then binds a serum protein called Factor B. Another serum protein, Factor D, then acts on bound Factor B converting the complex to C3b,Bb. C3b,Bb, like C4b,2b of the classical pathway, possesses C3/C5 convertase activity. After formation of C3b,Bb a stabilizing molecule known as properdin binds to the complex protecting it from degradation. Once the C3/C5 convertase is formed the classical and alternative pathways are identical. The end results of complement activation are opsonization of the pathogen (C3b), recruitment of inflammatory cells to the site of infection (C3a, C4a, & C5a), and direct killing of the pathogen by the MAC (C5b6789).

Like most effector systems the complement cascade has regulatory systems that prevent the response from becoming too exuberant or being inappropriately activated. If C3b binds to Factor B while free in serum rather than when bound to the

Figure 2.3. Immunoglobulin monomer: L = light chain; H= heavy chain, disulfide bonds, and c and v represent constant and variable regions respectively.

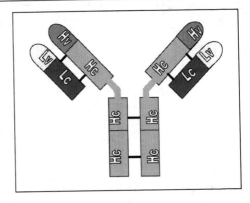

pathogen, the C3b,B complex is unstable and a protein called Factor H displaces Factor B. Factor I then acts on the C3b,H complex converting it to iC3b. Phagocytic cells possess receptors for iC3b which stimulate phagocytosis when they are bound much like the binding of C3b to its receptor, however iC3b does not appear to activate other components of the cascade so the response is not amplified.

Cells of the Innate Immune System

The cellular components of innate immunity consist of macrophages, neutrophils, Mast cells, eosinophils, and natural killer (NK) cells. The primary phagocytic cells, macrophages and neutrophils, have receptors on their surface for carbohydrate moieties such as lipopolysaccharide and mannose receptor that are not usually expressed by vertebrate cells. Phagocytic cells also have CR1 receptors for C3b and Fc receptors for antibodies on their surface. These carbohydrate and complement receptors allow phagocytic cells to attack antigens without assistance from the acquired immune system. These receptors allow the phagocytic cell to bind an antigen when complementary carbohydrate moieties or C3b is near one of the receptors (Fig. 2.4).

Carbohydrate and complement receptors bind in a zipper-like fashion until the antigen is completely enclosed by a membrane-bound vesicle called a phagosome. The phagosome then fuses with another vesicle filled with degradative enzymes known as a lysosome. This fusion forms a phagolysosome. Within the phagolysosome, the antigen is destroyed by myeloperoxidase, NADPH oxidase, cathepsins, and lysozyme. Early in the immune response neutrophils are the predominant phagocyte and they provide the primary defense against pyogenic bacteria. If the infection persists and the immune response continues, macrophages derived from circulating monocytes become more numerous. In addition to their role as phagocytes, macrophages secrete cytokine mediators, that modify the immune response. Neutrophils also secrete inflammatory mediators but their ability to do this appears limited by comparison to macrophages.

When a pathogen is too large to be phagocytized as in the case of certain parasitic worms, another cell type, the eosinophil, mediates the attack on the pathogen. Eosinophils express CR1 receptors and FcεRII, a receptor for IgE, on their surface. When the eosinophil encounters a pathogen opsonized by either C3b (Note: this can be mediated by innate immunity via the alternative pathway or acquired immunity via the classical pathway) or IgE its receptor for that particular opsonin is

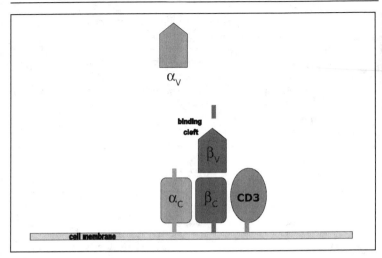

Figure 2.4. T cell receptor; v= variable region; c= constant region.

engaged and when enough receptors aggregate for crosslinking to occur the cell degranulates. The eosinophilic granules contain major basic protein, cationic protein, histaminase, and phospholipase D. Production of reactive oxygen intermediates is also upregulated in eosinophils by receptor crosslinking. By releasing these compounds the eosinophil mounts an extracellular attack against those pathogens that are too large to be engulfed.

C3a, C4a, and C5a are anaphylatoxins each capable of triggering mast cell degranulation and mediator synthesis to varying degrees. Mast cell granules contain preformed molecules of histamine, platelet activating factor (PAF), and interleukins 3, 4, 5 and 6. When appropriately stimulated, mast cells are also capable of producing leukotrienes, prostaglandins, and thromboxanes, which are all products of arachidonic acid metabolism. The result of mast cells releasing these mediators is vasodilation, increased vascular permeability, and recruitment of other inflammatory cells and proteins to the site of infection producing the cardinal signs of inflammation (rubor, calor, dolor, and tumor).

In addition to bacteria and parasites, the innate immune system is capable of defending against viral infection. Viruses lack the ability to reproduce independently. They must enter host cells and commandeer the replicative machinery to reproduce. When a virus infects a cell, the cell expresses certain high molecular weight glycoproteins on its surface distinguishing it from uninfected cells. Natural killer (NK) cells are large granular lymphocytes that possess receptors capable of recognizing these foreign glycoproteins and Fc receptors (FcR) for the Fc region of antibodies. When a NK cell binds to a cell by either of these receptors, it releases the contents of its granules into the extracellular space adjacent to the target cell. One of the proteins released, perforin, diffuses across the extracellular space and inserts into the target cell membrane. It then polymerizes with other perforin molecules to form a pore-like structure that allows water and solutes to pass freely between the intracellular and extracellular compartments. These porelike structures allow other granular contents such as endonucleases, granzymes, and TNF-β to enter the cell where they cause nuclear fragmentation and interfere with other vital functions resulting in apoptosis. When this process is initiated by binding of the Fc region to FcR, it is referred to as antibody dependent cell-mediated cytotoxicity (ADCC). Although perforin is capable of causing lysis under experimental conditions, its primary role physiologically seems to be providing a port of entry for these apoptosis-inducing proteins. After viruses enter a cell there is a lag phase before new viruses are produced; thus killing a few infected cells is beneficial to the host as a whole because the cell is destroyed before intact viruses capable of infecting neighboring cells are produced. NK cells act on cells infected with intracellular bacteria in a similar fashion and have also been demonstrated to have antitumor activity.

Acquired Immunity

The primary lymphoid organs are the bone marrow and thymus. All immune stem cells originate in the bone marrow. One population of stem cells remains in the bone marrow and differentiates into mature naive B cells while another population migrates to the thymus and differentiates into mature naive T cells. Each cell type is capable of responding to an antigen but has not yet received the stimulus required to elicit a reaction. Once these stem cells have differentiated into mature naive effector cells capable of responding to antigen, they leave the primary lymphoid organs and migrate throughout the body constantly circulating through the secondary lymphoid tissues. The secondary lymphoid organs consist of the spleen, lymph nodes, and MALT (mucosal-associated lymphoid tissue) that filter the blood, extracellular fluid, and substances crossing mucosal surfaces respectively for foreign antigens. Secondary lymphoid organs are sites where immune responses often begin because of their ability to concentrate antigen.

Acquired immunity is specific. The adaptive responses can be separated into two divisions, humoral and cell-mediated immunity. The essential functions of the two divisions are carried out primarily by B cells in the humoral response and T cells in the cell-mediated response. Antibodies produced by terminally differentiated B cells can discriminate between different antigens and T cells can discriminate between different processed antigens as well as between self and nonself proteins. The acquired system can also recruit the mechanisms of innate immunity to amplify its own responses. Once B cells have gone through the various precursor stages in the bone marrow, their specificity is determined by antibodies expressed on their surface that bind to complementary antigens. T cell specificity is determined by T cell receptors (TcR) on the cell surface. The TcR can only bind processed antigens that are presented to it in association with a major histocompatability complex (MHC) protein. In the thymus T cells go through numerous precursor stages and those cells that bind MHCs too strongly or with too low of an affinity are deleted. In addition, those cells that recognize processed self antigens are deleted as well. Deletion of these cells limits autoimmune reactions and is referred to as tolerance. Other tolerance mechanisms that occur outside the primary lymphoid organs are referred to as peripheral tolerance.

Functional Proteins of the Acquired Immune System

Antibodies

Antibodies are proteins that bind specifically to epitopes on antigen surfaces neutralizing them or opsonizing them for phagocytosis. Each antibody has a unique specifity and binds to a distinct epitope, however all antibodies possess the same basic structure and are referred to as immunoglobulins. Immunoglobulins are composed of two identical heavy chains and two identical light chains. There are two types of light chains, κ and λ, but each immunoglobulin contains only one type of light chain. The heavy chains are connected to one another and the light chains are connected to the heavy chains by disulfide bonds. The light chain has a constant (C) and variable region. The constant region is encoded by either a κ or λ gene. The variable region is encoded by two genes that are translated into a large V and a smaller J segment. There is a cluster of 70 or more V genes and 5 J genes. After transcription of the C, J, and V segments, recombinases link the mRNA together and the combined CJV mRNA is then translated into a complete light chain. Heavy chains are produced in a similar fashion except there is a D segment in the variable region between the V and J segments and there are five potential constant regions, γ, μ, α, δ, and ε (Fig. 2.3).

When the complete immunoglobulin molecule is digested by papain, one F_C fragment and two F_{Ab} fragments are produced. The F_{Ab} fragment composed of a light chain and the variable region of a heavy chain is the portion of the antibody that determines its specificity and binds to the antigen. Since each immunoglobulin monomer has two F_{Ab} fragments, its antigen binding valency is two. The F_C fragment consists of the constant region of the two heavy chains, and by binding to specific cell surface antibody receptors it determines the biological role of the antibody. Immunoglobulins are classified into IgG, IgM, IgA, IgD, and IgE isotypes based on the type of heavy chain each contains (γ, μ, α, δ, and ε respectively). These isotypes and receptors specific to each one determine the antibody's distribution within the body. Isotypes also differ with regards to half-life and ability to fix complement.

IgG is the most abundant antibody found in the serum and diffuses more readily into other body compartment than other isotypes. IgG is secreted as a monomer by terminally differentiated B cells called plasma cells. IgG can bind to and neutralize antigens as well as act as an opsonin. Macrophages possess high affinity IgG receptors, FcγRI, capable of binding free monomeric IgG. This receptor appears to be involved in macrophage mediated antibody dependent cell-mediated cytotoxicity (ADCC). The low affinity receptors, FcγRII and FcγRIII, are found on a wide variety of cells. These receptors bind to aggregated antibody (i.e., immune complexes). Although the strength of each individual receptor/antibody interaction is low, multiple antibodies binding provides high binding avidity. The concept of multiple antibodies binding at once is important because receptor occupancy alone does not initiate most cell functions. For some cell actions to occur, whether degranulation, phagocytosis, or extracellular killing, multiple receptors must be occupied allowing them to aggregate and interact. This process of interaction between aggregated receptors is referred to as crosslinking. FcγRII receptors are involved in phagocytosis by macrophages and neutrophils as well as degranulation of eosinophils while FcγRIII mediates NK cell ADCC and immune complex clearance by macrophages. In addition to the above abilities, IgG can fix complement but C1 requires binding of two IgG molecules for activation.

IgA is found primarily in secretory fluids where it defends surfaces such as the GU, GI, and respiratory tracts from infection. Plasma cells below the basement membranes of these epithelia secrete dimers of IgA. The IgA molecules are bound to one another within the plasma cell by a protein called the J-chain. The dimeric structure gives it an antigen binding valency of four. After secretion by a plasma cell, the IgA dimer binds to a poly-Ig receptor on the basal surface of the epithelial cell. This complex is then internalized and transported to the apical surface. At the apical surface of the epithelial cell, a portion of the poly-Ig receptor is cleaved from the complex and the IgA dimer with the remaining part of the poly-Ig receptor known as the secretory piece is released onto the mucosal surface. Once secreted, the secretory piece appears to protect the dimer from degradation. Dimeric IgA acts by binding to and neutralizing an antigen before it can invade the body. A small amount of monomeric IgA is produced and secreted into the tissue beneath the mucosa apparently functioning to neutralize those antigens that manage to evade the secreted IgA.

IgM is also known as macroglobulin because its circulating pentameric form has a very high molecular weight compared to other immunoglobulins. The pentamer, which is held intact by a J-chain similar to that found in IgA, gives it an antigen binding valency of ten and also allows one complete IgM molecule to fix complement. Each F_{Ab} fragment has low binding affinity; however because of its valency intact IgM has a high binding avidity. IgM is typically the first antibody type produced in the humoral response. Although its theoretical valence is ten, in reality it rarely binds more than five antigens due to steric hindrance. However, IgM is still very effective at agglutinating antigens and its ability to activate complement helps to destroy antigens. In addition to its pentameric form, monomeric IgM is also produced. These IgM monomers are produced with a hydrophobic region attached to the constant region that allows it to insert into the surface of mature naive B cells as a receptor. IgD is also expressed on mature naive B cells as a receptor, and it is almost exclusively found here. Little is known regarding the function of IgD.

IgE concentrations in the serum are very low. Mast cells express FcεRI, a very high affinity receptor for IgE. Monomeric IgE binds to these receptors on circulating mast cells. When enough antigen binds to the IgE on the mast cell surface it aggregates the FcεRI receptors close to one another allowing crosslinking. After crosslinking occurs, the mast cell degranulates. This is the basis for the type I hypersensitivity reaction.

T Cell Receptors (TcR)

Just as B cells express IgM and IgD antibodies on their surface to serve as receptors, T cells also express receptors on their surface each with specificity for a different antigen. However, unlike antibodies that recognize epitopes on intact antigens, T cell receptors can only recognize processed antigen bound to an MHC. The T cell receptor is composed of an α and a β subunit. The α subunit is analogous to the constant region of the immunoglobulin while the β subunit determines binding specificity of the receptor. The formation of the α and β chains, again similar to antibody production, involves a series of recombinations which allows a limited number of genes to produce a very large number of receptors each with a different antigen binding specificity. The TcR is associated on the T cell surface with a protein designated CD3.

Once the processed Ag-MHC complex binds to the cleft formed by $α_v$ and $β_v$ CD3 transduces the signal of binding to the cell interior initiating a series of reactions that elicit a specific effect dependent on the type of T cell.

MHC

The major histocompatability complex, known as human leukocyte antigen in humans, was first described in the context of transplant rejection. There are two types of MHC. MHC type I, corresponds to HLA-A, B, and C, and is found on all nucleated cells. The MHC-I is composed of a heavy peptide with three globular domains designated α_1, α_2, and α_3. A smaller peptide called β_2 microglobulin is noncovalently linked to the larger peptide completing the MHC-I complex. A binding cleft for the processed antigen is formed between the α_1, and α_2 domains. The peptides that bind to this cleft are processed in the cytosol.

MHC-II corresponds to HLA-DR, DQ, and DP in humans and is found primarily on antigen presenting cells (B cells, dendritic cells, and macrophages). MHC-II is composed of two peptides, an α and β chain. Each chain has two globular domains. The α_1 and β_1 domains form the binding cleft for MHC-II. Peptides that bind to this cleft are processed within intracellular vesicles.

As previously mentioned processed peptide must be bound to a MHC for the TcR to recognize the antigen. There is further restriction to TcR recognition of Ag. For T cells to interact with an antigen presented by MHC-I they must have a protein called CD8 associated with the TcR. CD8 interacts with MHC-I to stabilize the TcR and processed peptide allowing antigen recognition to occur. CD8 is associated with the TcR on cytotoxic T cells. Inflammatory and helper T cells have CD4 associated with their TcR rather than CD8. CD4 interacts with MHC-II stabilizing the TcR and peptide that again allows antigen recognition to occur.

Acquired Immune Responses

It would be highly inefficient for the body to produce thousands of immune cells specific to every antigen an individual might encounter in its lifetime. In addition to the energy expenditure maintaining this large number of immune cells would incur, it would also occupy a large amount of space. The immune system overcomes these limitations by clonal selection and expansion. According to the theory of clonal selection, once an immune cell encounters the antigen it is specific to it undergoes a series of divisions producing roughly a thousand immune cells within about three days. Each of these cells is capable of producing the appropriate response to that particular Ag whether a humoral or cell-mediated response is needed. The secondary lymphoid organs are positioned so that they efficiently filter the extracellular fluid, blood, and material that crosses mucosal surfaces for foreign antigens. Once an antigenic compound is sequestered within one of these peripheral lymphoid organs, the secondary lymphoid organs are arranged to efficiently present the antigen to initiate the appropriate immune response. In addition to the anatomic location and histologic organization of the secondary lymphoid organs, lymphocytes circulate continuously through the peripheral lymphoid tissues. They migrate from one lymph node to another then return to the venous system on their way to the spleen. If the cell does not encounter its specific antigen it will leave the vascular compartment to repeat the journey. In these ways the odds that the lymphocyte will encounter an antigen it is capable of responding to are increased. A subset of cells derived from clonal expansion do not become effector cells. Instead these cells become memory cells and migrate to the bone marrow until the antigen is encountered again. Upon reexposure to the antigen these memory cells allow for more rapid immune responses.

Cell-Mediated Immune Responses

Cytotoxic (CD8+) T Cell Activation

When a virus infects a dendritic cell the virus is processed within the cytosol. Processed peptide then binds to the cleft in MHC-I and this complex is expressed on the dendritic cell surface. Next a naive CD8+ T cell recognizes this peptide-MHC-I complex and binds to it via its TcR. CD28, another molecule associated with the TcR, then binds to B7 on the surface of the infected cell. This interaction between CD28 and B7 transduces a signal to the T cell interior causing it to produce IL-2. IL-2 then feedsback on the T cell causing it to proliferate. The progeny of the proliferative response differentiate into cytotoxic T cells in response to IL-2.

Once the completely differentiated armed cytotoxic CD8+ T cells are formed in the secondary lymphoid tissues, they leave to search the body for infected target cells expressing the appropriate peptide/MHC-I complex. Equipped with new cell surface adhesion molecules and chemokine receptors, the CD8+ T cells migrate to the infected sites. T cell binding to target cells is initially mediated by adhesion molecules that are not specifically elicited by the antigen. This transient nonspecific interaction allows the TcR to bind to the peptide/MHC-I complex. If the TcR and the peptide/MHC-I complex are complementary, the binding affinity of the nonspecific adhesion molecules is increased causing a longer lasting and stronger bond. This TcR/MHC-I interaction also causes vesicles containing effector molecules to polarize along the T cell surface adjacent to the target cell, and finally it causes the release of these effector molecules. These effector molecules include perforin 1 which creates holes in the target cell membrane and granzyme (a.k.a. fragmentins), a protease that damages proteins vital to cellular functions. In addition, IFN-γ is released which inhibits viral replication. Also, a cell bound effector on the T cell known as Fas ligand binds to Fas protein on the target cell inducing synthesis of proteins associated with apoptosis. Using these soluble and surface bound effector molecules the cytotoxic CD8+ T cell induces target cell death by apoptosis.

Inflammatory (Thl) Helper T Cell Responses

Phagocytic cells are capable of ingesting and destroying many different types of bacteria and parasites; however some pathogens have developed mechanisms for evading destruction once ingested. When infectious agents armed with these evasive strategies are encountered, small amounts of the pathogen are still degraded within phagolysomes and processed peptides are expressed on the surface of the phagocyte in association with MHC-II. When an inflammatory CD4+ helper T cell encounters the infected phagocyte, there is again nonspecific binding that allows the specific TcR and peptide/MHC-II complex to interact. This ligand and receptor interaction stimulates the Thl cell to release IL-2 that feedsback on the Thl cell causing proliferation and differentiation. The T cell progeny, armed inflammatory CD4+ T cells, now disperse to find target cells.

When the armed Thl cell encounters cells expressing the peptide/MHC-II complex, it is specific for the Thl cell it stimulated to release IFN-γ, TNF-β, IL-2, IL-3, GM-CSF, MCF, and MIF. Interferon-γ (IFN-γ) activates the macrophage so it more effectively kills ingested bacteria. Tumor necrosis factor β (TNF-β) kills chronically infected macrophages that can no longer kill the bacteria, which releases the bacteria so fresh macrophages can engulf and destroy them. IL-2 causes further Th1 cell proliferation. Granulocyte-monocyte colony stimulating factor (GM-CSF) and interleukin-3 (IL-3) stimulate the production of more macrophages by the bone marrow. Macrophage chemotactic factor (MCF) and migration inhibition factor (MIF) cause macrophages to accumulate at the site of infection. Through these numerous effectors the Th1 cell of the adaptive immune system makes the cells of the innate system more effective and amplifies the immune response.

Humoral Immune Responses

Humoral Helper (Th2) T Cell Responses

Mature naive B cells have IgM and IgD antibodies on their surface. When these B cells encounter an antigen complementary to their immunoglobulin receptor, they bind to the antigen and endocytose the receptor/antigen complex. The antigen is then degraded in the endosome. Next processed peptide binds to the cleft in the MHC-II and the peptide/MHC-II complex is expressed on the cell surface. When the B cell encounters a helper (Th2) T cell that is specific to the peptide/MHC-II complex, nonspecific cell adherence occurs until the TcR and peptide/MHC-II complex are brought into contact (Note: The processed peptide that the TcR recognizes may be different from the epitope the antibody receptor binds). Once this occurs, interaction between the TcR and the peptide/MHC-II complex tranduces signals to the interior of the helper T cell. This causes the helper T cell to express CD40 ligand on its surface that binds to CD40 on the B cell shifting the B cell into the cell cycle causing B cell proliferation. In addition to upregulating CD40 ligand on its surface, the Th2 cell secretes IL-4 that synergizes with the CD40 ligand to cause B cell clonal expansion. Next the Th2 cell secretes IL-5 and IL-6 that stimulate B cell differentiation. As the stimulated B cells leave the secondary lymphoid tissue they interact with follicular dendritic cells (FDC). CD23 on the FDC binds CD19 on the B cells. If a B cell receives no further stimulation, it develops into a plasma cell while if it contacts CD40 ligand it becomes a memory cell. The antibody initially produced is all of the IgM isotype, but dependent on the amounts of IL-4, IL-5, IFN-γ, and TGF-β (transforming growth factor β) secreted by the T cell the B cell undergoes isotype switching and becomes dedicated to producing either IgA, IgE, or one of the subclasses of IgG. Once the antibody is secreted by the plasma cells it circulates throughout the body. Antibody can neutralize the antigen preventing infection of more cells, opsinize antigen increasing the efficiency of phagocytosis, or become involved in ADCC.

If any of the costimulatory signals described in the above responses are lacking, the effector cell may become anergic. This prevents inappropriate activation of immune cells such as cells that are specific to self antigens. This is referred to as peripheral tolerance.

Hypersensitivity Reactions

Hypersensitivity reactions occur when the body mounts an excessive immune response to a typically innocuous antigen resulting in tissue damage. Hypersensitivity reactions require a sensitizing exposure so the excessive response does not occur with the initial exposure. There are four classic hypersensitivity reactions. Types I, II, and III occur shortly after reexposure to the antigen and are therefore called immediate hypersensitivity reactions. Type IV reactions on the other hand occur days after reexposure to the incitng antigen so they are referred to as delayed-type hypersensitivity reactions. Once sensitized the reactions often worsen in severity with each subsequent exposure due to increased antibody or T cell production.

Type I reactions occur within seconds or minutes after exposure to the antigen. Antigens that provoke this type of reaction are usually small, soluble proteins capable of eliciting an immune response at low doses (examples: pollen, pet dander, penicillin). After the initial exposure to the antigen, Th2 helper cells direct B cells to become IgE producing plasma cells. Monomers of IgE antibody specific to the inciting antigen then bind to FcεRI on mast cells with high affinity. Upon reexposure to the antigen the IgE on the mast cell binds to the antigen causing aggregation and subsequent crosslinking of the FcεRI resulting in degranulation. Mast cell degranulation releases histamine, leukotrienes, and cytokines that cause smooth muscle contraction, capillary dilatation, and increased vascular permeability. The severity of the clinical presentation is dependent on the amount of antigen innoculum, route of antigen exposure, and the amount of IgE produced. Cutaneous exposure to limited amounts results in a flare and wheal. Intravenous innoculation may result in systemic anaphylaxis that should be treated with epinephrine. Inhalation of the antigen may cause allergic rhinitis or asthma. Patients with reactions that are not life threatening can be treated with antihistamines for symptom relief, and the later sequelae can be treated with corticosteroids. Patients can be desensitized to the antigen by controlled exposure that shifts the antibody produced from an IgE to IgG and IgA decreasing the likelihood that receptor crosslinking and subsequent degranulation will occur since the IgG and IgA will compete with IgE making FcεR aggregation less likely.

Type II reactions occur a few hours after exposure to the antigen. Antigens that cause this type of reaction bind to cell surfaces. After initial exposure to the antigen, IgG antibodies are formed. Upon reexposure to the antigen, memory cells in the bone marrow become activated and a large amount of antigen specific IgG is produced. This IgG then binds to antigen on the surface of host cells. Some IgG then fixes complement initiating the complement cascade and the Fc portion of some of the IgG binds to Fc receptors on macrophages and neutrophils. Activation of the complement cascade and phagocytic cells results in destruction of the cells to which the antigen is adherent. RBCs and platelets are typically the targets of this type of reaction.

Type III reactions occur several hours after antigen exposure. The antigens responsible for these reactions are usually soluble. When there is antibody in excess compared to antigen, insoluble immune complexes form while the complexes are soluble if there is antigen excess. Upon repeat exposure to the antigen, memory cells are activated resulting in production of large amounts of antibody that then form complexes with these antigens. These small complexes then become deposited in various tissues. Soluble complexes are widely distributed while particulate complexes tend to become deposited near the site of antigen entry. Deposited complexes activate complement and phagocytic cells. Since phagocytes cannot phagocytize the deposited complexes, they release their granular contents rather than engulfing the particles and tissue damage results. Clinical presentation is dependent on the route of exposure to the antigen. If the antigen is injected into the skin, IgG diffuses into the soft tissue, and immune complexes are formed which activate complement resulting in the release of vasoactive and chemotactic mediators. Release of factors such as C5a allows immune cells and fluid to enter the area from the vasculature. Complement and granular contents from immune cells cause tissue damage. This is referred to as the Arthus reaction. Inhalation of an antigen can cause an intrapulmonary Arthus reaction with the classic example being Farmer's lung when an individual inhales actinomycetes from mouldy hay. Serum sickness occurs when a large dose of antigen is given intravenously with horse serum antivenin being the classic example. After antigen exposure the immune complexes that form are dispersed to joints, renal glomeruli, and vessel walls as well as other sites. About eight days after the injection the patient becomes febrile, lymph nodes and joints swell, an urticarial rash appears, and the patient develops proteinuria.

Type IV hypersensitivity also referred to as contact hypersensitivity occurs one to three days after antigen exposure. In contrast to the immediate hypersensitivity reactions that are mediated by the formation of antibodies (i.e., Th2 mediated response), delayed hypersensitivity is mediated by Th1 helper cells. Antigens that elicit this type of response are capable of forming stable complexes with host proteins making them antigenic. The altered host protein is then endocytosed, processed by the cell, and expressed on the cell surface with MHC-II. Reexposure to the antigen activates previously sensitized Th1 (CD4) cells that enter the site where the antigen is located. Antigen presenting cells then present the processed antigen/MHC-II complex to the helper T cells that stimulates these cells to release various cytokine mediators resulting in fluid and immune cell influx into the area. Examples of this type of reaction include the response to gliadin (celiac disease), pentadecacatechol (poison ivy), and tuberculin (Mantoux reaction/PPD skin test). Alternatively, type IV reactions can be mediated by CD8 T cells recognizing processed antigen in association with MHC-I that activates their cytotoxic action.

Autoimmune Disease and Immunodeficiency

In addition to the excessive responses that produce hypersensitivity, inappropriate and inadequate responses occur resulting in either autoimmune disease or immunodeficiency. The occurrence of either type of reaction is infrequent, but they are worth mentioning because they show how the immune system should normally function.

Autoimmune diseases occur when the adaptive immune system mounts a sustained response against self antigens resulting in long term tissue damage. Autoantibodies can cause destruction as in the case of autoimmune hemolytic anemia. Autoantibodies can also be stimulatory. In Grave's disease antibodies to the thyroid stimulating hormone receptor are formed which stimulate thyroid hormone release resulting in thyrotoxicosis. Autoantibodies can also cause hypofunction of an organ. In myasthenia gravis antibodies form against the acetylcholine receptor on motor end plates causing the receptor to become depleted which results in muscle weakness. In systemic autoimmune diseases such as systemic lupus erythematosus and rheumatoid arthritis a myriad of autoantibodies form as well as altered cell mediated immunity result in tissue damage characteristic of each of these diseases.

In immunodeficiency the body fails to respond appropriately to eradicate an infection. In chronic granulomatous disease the cytochrome b oxidase system of phagocytic cells is defective so they are incapable of forming reactive oxygen intermediates to destroy engulfed pathogens. These patients have chronic infections with *Staphylococcus aureus*, candida, and aspergillus being the organisms that trouble these patients most. In hereditary angioedema, individuals lack activated C1 inhibitor and this results in recurrent episodes of acute edema mediated by C2a. Also patients

Table 2.1. Cytokines and their actions

Cytokine	Primary Sources	Actions
IL-1	Macrophages and epithelial cells	Pyrogen, T cell activator, and macrophage activator
IL-2	T cells	T cell growth factor
IL-4	T cells and mast cells	B cell activator and induces isotype switching to IgE
IL-5	T cells and mast cells	Eosinophil growth factor
IL-6	T cells and macrophages	T cell and B cell growth and differentiation; stimulates production of acute phase proteins in the liver
IL-8	Macrophages	Chemotactic agent for neutrophils
IL-12	Macrophages	Activates NK cells and stimulates Th1 differentiation of CD4 T cells
IL-13	T cells	B cell growth and differentiation
IFN-γ	T cells and NK cells	Activates macrophages and increases MHC expression
IFN-α	Leukocytes	Inhibits viral replication and increases MHC-I expression
IFN-β	Fibroblasts	Inhibits viral replication and increases MHC-I expression

deficient in any component of the MAC, the terminal component of the complement system, have been shown to be prone to Neisserial infections. In Bruton's γ-globulinemia, an X-linked syndrome, individuals produce limited amounts of immunoglobulin because of defective heavy chain gene rearrangement. These individuals have recurrent infections with staphylococci, streptococci, and hemophilus. Individuals with DiGeorge syndrome lack sufficient numbers T cells because their thymus failed to develop appropriately from the third pharyngeal pouches. These individuals have no cell-mediated responses and humoral responses are depressed. Affected individuals also lack parathyroid glands and have cardiovascular abnormalities. The most pervasive immunodeficiency is severe combined immunodeficiency (SCID). In these individuals defective recombinase enzymes prevent formation of T- and B cell receptors thus they are very susceptible to all infectious agents. These patients can be treated with a bone marrow transplant if a compatible donor is available. Despite the fact that these disorders are very uncommon they do illustrate the importance of an intact and appropriately functioning immune system.

Conclusion

The immune system is a complex system and the breadth of our knowledge is constantly expanding. The role of soluble mediators as key components of the system is becoming clearer. Where appropriate the role of specific mediators has been described in immune responses in the body of the chapter. For the sake of completeness a table describing the action of the more important and well known mediators is included (Table 2.1). The immune system is redundant and has multiple feedback loops. In conclusion, the immune system is designed to protect the body from invasion by foreign substances and maintain the health of the individual.

References

1. Janeway CA, Travers P. Immunobiology: The Immune System in Health and Disease. London: Current Biology Ltd., 1994.
2. Roitt I. Essential Immunology. 8th ed. Cambridge, MA: Blackwell Scientific Publications, 1994.
3. Delves PJ, Roitt I. The Immune System, first of two parts. N Engl J Med Adv Immunol 2000; 343:37-49.
4. Delves PJ, Roitt I. The Immune System, second of two parts. N Engl J Med Adv Immunol 2000; 343:108-117.

CHAPTER 3

Hemostasis

William R. Wrightson

Management of Coagulation in the Surgical Patient

History

Virchow in 1856 described the famous triad:
1. Stasis
2. Endothelial damage
3. Hypercoaguable states

The basis of the coagulation system is based on the coagulation cascade. The end points of this cascade include the formation of thrombin and fibrin. Throughout this system there can be defects in the multiple enzymes or extrinsic factors contributing to its dysfunction.

There are several interrelated stages in the formation of a clot:
- Vasoconstriction
- Platelet aggregation
- Intrinsic pathway
- Extrinsic Pathway

The initial response in injury is a transient vasoconstriction which is linked to platelet plug and fibrin formation. Vasoconstriction is mediated through intrinsic mechanisms and various vasoactive agents (thromboxane A2 and serotonin) released during platelet aggregation.

Platelets

Normal count is 150-400k with typical life span of 7 days. Primary hemostasis is associated with platelet adherence to subendothelial collagen of disrupted tissue. This requires von Willebrand factor (vWF) to work. An aggregate of platelets that subsequently forms with further aggregation mediated by ADP, thromboxane and serotonin. Opposing mediators include prostacyclin and endothelium-derived releasing factor. These are vasodilators and inhibit aggregation.

Coagulation Cascade

Two pathways make up the system: intrinsic and extrinsic. The pathways are named based on in vitro experiments that used "extrinsic" source of tissue thromboplastin (usually brain tissue) to induce the cascade. The second system was initiated in the test tube without and external stimulus (i.e., the "intrinsic" system).

The intrinsic pathway is initiated by exposure of coagulation factors to subendothelial collagen. Tissue factor combined with various other factors to converge on factor X which cleaves prothrombin to thrombin. The extrinsic pathway is activated by tissue factors or glycoproteins.

Current Concepts in General Surgery: A Resident Review, edited by William R. Wrightson. ©2006 Landes Bioscience.

Table 3.1. Coagulation factors

Factor	Name
I	Fibrinogen
II	Prothrombin
III	Tissue thromboplastin
IV	Calcium
V	Proaccelerin
VI	
VII	Proconnectin
VIII	Antihemophilic factor
IX	Christmas factor
X	Stuart-Prower factor
XI	Plasma thromboplastin
XII	Hageman factor
XIII	Fibrin stabilizing factor

All of the coagulation factors are synthesized in the liver except thromboplastin, factor VIII and calcium. Factors II, VII, XI, X, protein C and S are vitamin K dependent enzymes and therefore effected by coumadin (Table 3.1).

Calcium plays an integral role in the cascade and is required as a cofactor in many of the steps. Hypocalcemia can lead to coagulopathy and should be monitored in patients with coagulation defects especially following significant transfusions (Table 3.2).

Fibrinolysis

There is a delicate balance between formation and lysis of clot. Lysis of fibrin deposits is mediated by antithrombin III, protein C and S and plasmin. Antithrombin II as the name suggests blocks thrombin. When combined with heparin it also blocks factors XII, XI, IX and X (intrinsic pathway).

Diagnosis of Coagulation Disorders

Tests used to measure fibrinolysis include fibrin degradation products (FDB), fibrinogen, d-dimer (Table 3.3).

Table 3.2. Coagulation cascade

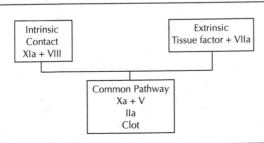

Table 3.3. Tests of coagulation

Test	Purpose
Prothrombin time (PT)	Extrinsic system
Activated partial prothrombin time (PTT)	Intrinsic system
Thrombin time (TT)	Detect abnormalities in fibrinogen and anticoagulants
Template bleeding time (BT)	Assess interaction between platelets; Usually related to platelet dysfunction

Defects in Hemostasis

Hemophilia (Factor VII)

This is a sex-linked recessive disorder resulting in a failure to synthesize normal factor VII. Incidence is 1 in 10000. There are many degrees of severity of the disease with as little as 5% of normal factor VIII controlling bleeding. Patients may present early in life with bleeding complications or later in life secondary to sustained trauma. Patients may also present with hemarthrosis, retroperitoneal bleeds or intestinal hematomas and obstruction.

Management

The half life of factor VII is 8-12 hours. In most circumstances hemostasis can be maintained with as little as 3-5% of normal levels, however, active bleeding may require levels of closer to 30%. Cryoprecipitate is given as 1 unit/kg to raise the level by 1 percent. Patients undergoing major procedures should have levels raised to 30% for 10 days following the procedure. Even minor cases should achieve these levels.

Christmas Disease

This is a factor IX deficiency and is X-linked recessive. Severe disease results with less than 1% normal activity. All patients require substitution therapy when surgery is performed.

von Willebrand Disease

This factor assists in the binding of platelets to endothelium. The most important receptor involved in its function is platelet glycoprotein-Ib. This disease occurs in approximately 1 in 1000 people and is inherited as an autosomal dominant trait. It is associated with abnormal vWF and decreased stabilization of factor VIII with a prolonged bleeding time. Ristocetin fails to cause platelet aggregation in 70% of patients with the disease. Treatment with cryoprecipitate day before surgery and continue for 4 days.

Platelet Dysfunction

Thrombocytopenia is probably the most common defect encountered in the surgical population. Platelet disorders can be qualitative or quantitative. Quantitative deficits can be the result of consumption, decreased production, increased destruction or sequestration.

Table 3.4. Etiologies of thrombocytopenia

Quantitative Defect

Drugs	Thiazides
	H2 blockers (cimetadine)
	Heparin
	Penicillin
	Cephalosporins
	Sulfonamides
	Alcohol
Viral	CMV
	EBV
	Herpes simplex
Immune	Sepsis
	Drugs (heparin, quinidine, sulfonamides)
	ITP
	SLE

Qualitative Defect

Drugs	Aspirin
	NSAIDS
	Corticosteroids
	Antibiotics (Penicillin, cephlasporin, etc)
	Antihistamines
	Alpha/beta blockers
	DextranSurosemide
	Heparin
	Lidocaine
	TCA
Other	Uremia
	Hypothyroidism
	DIC

Counts >50 k do not require treatment. One unit of platelets will increase count by 10k. Spontaneous bleeding <20 k treat. Surgery safely performed as long as count is >50k.

Heparin-Induced Thrombocytopenia

Reported in 0.6% of patients receiving heparin. Likely immune mediated. Counts decrease to 4-15 days after initial treatment. Treat by stopping the heparin. If patient requires anticoagulation alternative therapy includes Refludan® (Table 3.4).

Diffuse Intravascular Coagulation (DIC)

Introduction of thromboplastic agents into circulation. Associated with extracorporeal circulation, sepsis, lymphoma and a prolongation in TT, PTT, PT. Diagnosed by a reduced fibrinogen, increased FDP, decreased platelets. Management consists of controlling the underlying problem. Replace with FFP, cryoprecipitate, and platelets if the patient is actively bleeding. In most instances the DIC leads to thrombotic complications. Heparin combined with component therapy to control both the loss of coagulation factors and thrombotic tendency. Epsilon aminocaproic acid (EACA) is an inhibitor of fibrinolysis and is prone to induce thrombosis. Therefore any patient receiving this should receive heparin.

Head Injury

The brain is the richest source of tissue thromboplastin. Closed head injury may be associated with coagulopathy from this release. Treatment involves replacement of clotting factors with FFP and close monitoring for DIC.

Hypercoaguable States

There are several protein deficiencies that affect the normal process of coagulation and can lead to a hypercoaguable state. Most of these are serine protease inhibitors.

Antithrombin III Deficiency (ATIII)

ATIII inhibits thrombin in vivo and in accelerated 1000-10,000 fold in the presence of heparin. Inherited as an AD disorder. Patients with this disorder have a 50% risk of thrombotic events in their lifetime. They also are not sensitive to heparin.

Protein C or S Deficiency

Both are serine proteases and serve to inactivate factor Va and VIIIa with protein S serving as a cofactor for APC (activated protein C).

Factor V Leiden

This results from a mutant factor V. This mutant when activated is resistant to APC. Approximately 5% of the US population is heterozygous for this defect.

Correction of Anticoagulation in Surgical Patients

Patients may be anticoagulated for many reasons. Ultimately, recommendations for reversal of anticoagulation for both major or minor surgical procedures are as indicated in the following section.

If a patient is therapeutic with their coumadin therapy, four scheduled doses should be withheld prior to surgery to allow the INR to drop to ≥1.5. The INR should be measured the day before surgery and further corrected if necessary. If heparin is required it should not be restarted until 12 hours after surgery and delayed if there is any evidence of bleeding.

Only patients with a history of acute thromboembolism (<1 month) need continuous anticoagulation with heparin pre- and postoperatively. Patients with heart valves and atrial fibrillation do not require preoperative heparin therapy. The risk for embolic events in the short term is small ranging between 0.3% to 1.1% at two weeks (Table 3.5).

Table 3.5. Management of anticoagulated patients

Indication		Before Surgery	After Surgery
Acute venous thromboembolism			
	1 month	IV Heparin	IV Heparin
	2-3 months	None	IV Heparin
Recurrent venous thromboembolism		None	SC Heparin
Acute arterial thromboembolism	(1 month)	IV Heparin	IV Heparin
Mechanical heart valve		None	SC Heparin
Atrial fibrillation		None	SC Heparin

Table 3.6. Medications affecting coagulation

Class	Example	Effects	Comments
Aspirin	ASA	Platelet inhibitor, irreversible. 7 days	DDAVP has some success in reversal. Platelet transfusion
Antiplatelet agents	Clopidogrel (Plavix), cilostazol (Pletal)	Platelet aggregation inhibitor, inhibits cellular phosphodiesterase	Discontinue 7 days before major surgery
Heparin	Heparin	Binds antithrombin III	Stop 4 hours prior to surgery. Must follow PTT. Watch for HIT
Warfarin	Coumadin	Inhibits vitamin K dependent enzymes	Need 4 days to reverse
Lepirudin	Refludan	Directly inhibits thrombin	Heparin substitute for HIT
DDAVP	Desmopressin	Increases factor VII levels, increases release of vWF	Transient response
Aprotinin	Trasylol	Serine protease inhibitor; inhibits plasmin, tPA and serum urokinase plasminogin activator	Decreases bleeding after cardiopulmonary bypass

3

Medications
(Table 3.6)

References
1.   Kearon C, Hirsh J. Current concepts: Management of anticoagulation before and after elective surgery. NEJM 1997; 336:1506-1511.
2.   Spiess BD, Counts RB, Gould SA. Perioperative Transfusion Medicine. 1st ed. Baltimore: Williams and Wilkins, 1998.

Fibrin Sealants and Hemostatic Agents in Surgery

Fibrin sealants have become a popular option for hemostatic agents. Their use includes hemostasis, wound closure and fistula closure. Unlike cyanoacrylates that have been used in dental procedures, there is no inflammation and these compounds are biocompatible.[1] The sealant is reabsorbed through native proteolytic systems.

Fibrin Glue Components

These products are produced as two separate solutions.
1. Lypholized pooled human fibrinogen/factor XIII concentrate (reconstituted with aprotinin)
2. Bovine thrombin (reconstituted with 40 mM $CaCl_2$)

The combination of these two components then can form a stable clot. The essential components to fibrin glue are fibrinogen, thrombin and calcium. Factor XIII is also required but endogenous factor is usually sufficient to induce clot formation.

Table 3.7. Transfusion transmitted disease incidence

Transmitted Disease	Incidence of Transmission
Hepatitis C	1:100,000
Hepatitis B	1:200,000
HIV	1:450,000

Viral Transmission

As noted, fibrinogen is an essential factor in the glue and can be obtained from screened donor plasma minimizing the risk of hepatitis and HIV. Both fibrinogen and thrombin are subjected to a variety of treatments to minimize risk of HIV, hepatitis B and C, HTLV, EBV and CMV. Parvovirus B19 have been shown to be transmitted; however most patients have antibodies to this making any infection subclinical. Viral inactivation has been shown to be greater that 9 $\log_{10}$ to 19 $\log_{10}$ for HIV-1, HSV-1 in fibrinogen, factor XIII and thrombin.[1] While the risk is low, it has not been eliminated (Table 3.7).

Physiology

Once these components are combined they form a stable physiologic fibrin clot. Is essence, it jumps to the final stages of the coagulation cascade to combine thrombin and fibrin to obtain coagulation. The strength of the clot is a function of the fibrinogen concentration with the rate of clot formation a function of the thrombin concentration.

Hemostasis

Fibrin glue has been used successfully in cardiothoracic surgery for many years. In patients undergoing reoperation for continued bleeding, fibrin glue produced hemostasis in 92% of patients where traditional methods were successful in only 12%.[3]

In many instances the raw cut surfaces following resection cannot be controlled with conventional techniques. Fibrin glue has been used in hepatic surgery with some success. It has also been suggested that fibrin glue helps to seal severed bile ducts following resection. This results in decreased blood loss and reduced biliary leakage.[2]

Tissue Sealant

Application in pancreatic surgery has shown a 25% decrease in the incidence of fistula formation when applied to the pancreatic stump intraoperatively. Other studies have not been as optimistic and have not found a significant difference found between treatment groups.[2]

Bronchopleural fistula is a common complication following pulmonary resection. Fibrin sealant has been shown to decrease the incidence of pneumothorax following percutaneous lung biopsy. Current animal studies have shown some benefit of the sealant in reducing pulmonary air leaks.[2]

Other Hemostatic Agents

There are many other agents available to control minor bleeding.

Surgicel®. This is oxidized cellulose that reacts with blood to form a clot. It will absorb in 2-4 weeks. Surgicel® also has minimal bactericidal effect but delays wound healing.

Gelfoam®. This acts to absorb and promote imitation of coagulation. It is typically combined with thrombin.

Avitene®. This is a microcrystalline collagen that adheres to platelets to stimulate clotting.

References

1. Jackson MR. Fibrin sealants in surgical practice: An overview. Am J Surgery 2001; 182:1S-7S.
2. Morikawa T. Tissue sealing. Am J Surgery 2001; 182:29S-35S.
3. Rousou J, Gonzalez-Lavin L, Cosgrove 0 et al. Randomized clinical trial of fibrin sealant in patients undergoing resternotomy or reoperation after cardiac operations. J Thoracic Cardiothoracic Surgery 1989; 97:194-203.

Suggested Reading

Jackson MR. Developing surgical techniques: The role of fibrin sealants. Am J Surgery 2001; Supplement.

Blood Products

Packed Red Blood Cells (PRBC)

Each unit of packed red blood cells (PRBC) contains approximately 450 mL of red blood cells with a small amount of plasma and anticoagulant. The hematocrit is usually around 60%. PRBC can be stored for 42 days. These concentrated solutions have most of the plasma and its constituents removed and therefore do not contain platelets, albumin or clotting factors.

Approximately 90% of transfusions are either allergic or febrile. This is most commonly due to antibodies directed against donor leukocytes; however it can be a sign of a transfusion reaction. Leukocytes in donor blood can be removed during transfusion with cotton or synthetic fiber filter. Some have suggested premedication with diphenhydramine and acetominiphen to diminish this reaction.

Complications of Blood Transfusion

Transfusion reaction is the result of hemolysis from improperly typed and crossed blood or clerical error (80%). It occurs in 1 of 6000 transfusions. Manifestations of hemolysis occur after 50 mL or less of blood has been transfused. The patient develops fever and chills. The symptoms may progress to headache, back pain, substernal chest pain, dypsnea and hypotension.

Management consists of discontinuing the blood transfusion and sending the blood for repeat type and cross. Resuscitation is initiated with maintenance of a high urine output (see Table 3.1).

Platelets

Each unit of platelets contains 5×10^{10} platelets in 60 mL of plasma. They are stored for up to 5 days. They must be stored at room temperature to maintain their activity (~70%). Each unit of pooled platelets will typically increase the platelet count by 10,000. Platelets are given to correct nonimmune thrombocytopenia. Patients with immune thrombocytopenia (ITP) do not respond with increased counts following administration of platelets. If these transfusions are needed to support surgery for ITP splenectomy, then they should not be administered until the splenic hilum is clamped.

Because they are stored at room temperature and cannot be shaken once pooled, they must be used within 4 hours of preparation.

Fresh Frozen Plasma (FFP)

FFP is produced from the separation of plasma from donated whole blood. It is subsequently stored at -18°C and has a shelf life of up to 1 year. Each unit contains about 400 mg of fibrinogen and 1 unit activity of each of the clotting factors. The most labile clotting factors (V and VIII) may be diminished proportional to shelf life. Once thawed, it can be stored for up to 24 hours before it must be discarded.

Its primary uses include:
1. Replacement of clotting factors for deficiency
2. Massive transfusion bleeding diathesis
3. Antithrombin III deficiency
4. Reversal of warfarin effect

Cryoprecipitate

This is a plasma concentrate of primarly factor VIII and fibrinogen suspended in 10 mL of plasma. The concentration of factor VIII is 100 units and fibrinogen 250 units (1 unit is the amount in 1mL of FFP).

Erythropoesis

Erythropoetin (Procrit®) has been suggested as a means of increasing the mean hemoglobin concentration by stimulating erythropoesis. Multiple randomized studies have showm that perioperative use of erythropoietin does increase Hgb levels compared to controls. One such study showed a 50% decrease in the need for postoperative blood transfusions following cardiac surgery.[1]

Red Blood Cell Substitutes

Polymerized Pyridoxylated Hemoglobin. This consists of lysed RBCs that have had all stromal elements removed and subsequently polymerized to decrease oncotic pressure.[2] This has been successfully evaluated in baboons and is under clinical investigation in humans.

Perfluorocarbon. These possess a unique ability to carry and exchange gas. They consist of a carbon skeleton saturated with fluorine. The soluability of oxygen is up to 20 times that of water. An emulsified solution is required to solubilize this in plasma. This process limits the carrying capacity to approximately 6 times plasma. Current studies are underway with second generation perfluorocarbons.[2]

References

1. Spiess BD, Counts RB, Gould SA et al. Perioperative Transfusion Medicine. 1st ed. Baltimore: Williams and Wilkins, 1998.
2. Gould S. Human polymerized hemoglobin first therapeutic use as a blood substitute in trauma and surgery. American College of Surgeons. New Orleans, LA: 81st Clinical Congress, October 22-27,1995.

CHAPTER 4

Anesthesia
Brian R. Davis

Physiology and Mechanism of Anesthetic Agents

Background

Anesthesia is a state of reversible loss of awareness and reflex reactions to noxious stimuli. The many diverse agents of anesthesia interact with neural cell membranes on a molecular basis to affect transmission of pain stimuli. These agents primarily interact with ion channels responsible for propagating action potentials.

Physiology

Intravenous Anesthetic

Intravenous anesthetic acts primarily at the GABA receptors producing hyperpolarizaztion of the nerve cell membrane. Benzodiazepenes, thiopental, propofol and etomidate bind modulator sites of the GABA A subtype receptors enhancing the inhibitory effect of these neurons. Unlike other intravenous anesthetic, ketamine depresses CNS activity by suppressing excitatory signals mediated by L-glutamate on NMDA receptor channels.

Inhalational Anesthetics

The molecular physiology of inhalational anesthetics is not completely understood with several hypotheses explaining anesthetic and channel interactions. The volume expansion hypothesis proposes that the anesthetics enter lipid membranes and create lateral pressure on channels preventing the influx of ions. The membrane fluidization hypothesis proposes that the anesthetic increases membrane protein motility and disrupts the arrangement of membrane lipid molecules causing ionic channels to lose their structural support and function. The protein interaction hypothesis proposes a direct action of inhalational anesthetics on amphophillic channel proteins thus modulating their gating mechanisms (Table 4.1).

Local Anesthetics

Local anesthetics also are felt to interact with membrane receptors in multiple different ways. The membrane receptor hypothesis states that local anesthetics enter the axon as uncharged bases, gain a charge forming acids which then bind receptor sites in sodium channels that prevent depolarization and impulse propagation. The membrane expansion hypothesis states that uncharged molecules such as benzocaine dissolve in the lipid matrix of the cell membrane allowing these membranes to expand and compress sodium channels thus blocking the passage of sodium ions, preventing subsequent depolarization. The surface charge hypothesis postulates that

Current Concepts in General Surgery: A Resident Review, edited by William R. Wrightson. ©2006 Landes Bioscience.

Table 4.1. Inhalational anesthetic characteristics

Agent	Minimum Alveolar Concentration	Comments
Nitrous oxide	105	Analgesic with little respiratory or cardiac depression.
Halothane	0.75	Slow uptake and elimination; cardiac depression; uterine relaxation
Isoflurane	1.15	Muscle relaxation; stable cardiac; neurosurgical application
Enflurane	1.68	Muscle relaxation; seizure activity

positively charged anesthetics bind to the axonal membrane making the external surface more positive than the internal surface thus hyperpolarizing the axonal membrane.

Muscle Relaxants

Muscle relaxants act directly on the motor endplate acetylcholine channels.

Nondepolarizing muscle relaxants are acetylcholine analogues that are competitive inhibitors binding reversibly with ligand gated channels preventing their activation by acetylcholine. As the concentration of relaxant molecules increases the population of functioning ion channels declines. When sufficient numbers of receptors are occupied, action potentials fail to propagate producing muscle relaxation. Nondepolarizing muscle relaxants also block access of acetylcholine to prejunctional receptors interfering with the mobilization of this neurotransmitter.

Depolarizing muscle relaxants have agonist activities at acetylcholine receptors causing prolonged depolarization of the motor end plate, interrupting neuromuscular signal transmission.

Opiates

Opiates act on the body's natural anesthetic pathways as analogues of the enkephalins and the endorphins at opioid receptors. There are five species of opioid receptor including the mu, delt, kappa, sigma and epsilon receptors. The opioids act primarily as agonists on the mu receptors producing symptoms of analgesia and euphoria.

General Anesthesia

Background

General anesthesia is used for operative procedures requiring unconscious sedation and muscle relaxation involving the torso and head or in those procedures to the perineum and extremities not amenable to regional anesthesia. There are four phases of general anesthesia including:

1. Induction
2. Maintenance
3. Emergence
4. Recovery

Induction is produced initially by administering an agent to cause sedation such as thiopental, midazolam, or propofol until loss of consciousness occurs noted by the loss of a blink reflex.

Induction also occurs with inhalational agents, often through combinations of nitrous oxide with halothane or enflourane. Loss of consciousness with these agents is noted by a loss of nystagmus and divergent gaze, midsized pupils, and a regular deep breathing pattern. Once unconscious, the patient is given a dose of muscle relaxant and intubated.

Maintenance of the unconscious state can be performed through the use of inhalational agents or with intravenous agents. During procedures on the head or extremities the patient is often allowed spontaneous ventilation with the use of a mixture of nitrous oxide, oxygen and inhalational agent to maintain unconsciousness. The neuromuscular blocker used for maintenance of anesthesia is determined by the length of relaxation required for the procedure.

Emergence from anesthesia in a patient who is breathing spontaneously occurs as soon as the inhalational agent is discontinued. Emergence occurs when muscle relaxation reverses through the self limiting effects of depolarizing agents or the reversal of nondepolarizing agents with the administration of an anticholinesterase.

Recovery from anesthesia involves adequate analgesia, the residual sedation from anesthetic agents, and control of postoperative nausea and vomiting.

Anesthetic Agents Used in Induction and Maintenance

Intravenous Agents

Most induction is performed with one or a combination of intravenous agents:
- Barbituate (thiopental)
- Benzodiazepine (midazolam)
- Propofol
- Ketamine
- Etomidate

Thiopental is a common barbituate used for induction, having a rapid onset of action and short duration. Thiopental decreases intracranial pressure and the cerebral metabolic rate making it useful for initiating a barbituate coma in severe head trauma. Thiopental depresses the myocardium, causes peripheral vascular pooling, and airway irritability leading to laryngospasm and bronchospasm thus limiting usefulness in trauma and emergency intubations.

Midazolam and diazepam are useful for induction in trauma and with cardiac surgery patients since they have modest cardiovascular side effects and additionally reduce cerebral metabolic rate and blood flow. Midazolam is more frequently used than diazepam because of its faster hepatic extraction and lack of active metabolites causing little residual sedative effect after surgery. Reversal of benzodiazepines can be accomplished with flumezanil which acts as a competitive antagonist at GABA receptors.

Propofol is commonly used as an induction agent because of its rapid onset of action, antiemetic properties and extrahepatic metabolism. Propofol has a greater cardiac depressive effect than the benzodiazepines and barbiturates but does not cause bronchospasm making it the agent of choice in patients requiring airway instrumentation.

Etomidate has minimal adverse effects on the respiratory and circulatory systems making it useful for patients with hypovolemia, cardiac disease, and asthma.

Side effects of etomidate include myoclonic jerks, hiccups and postoperative nausea and emesis limit its use.

Ketamine is useful in hypovolemic patients through its stimulatory effects on the cardiovascular system and its ability to increase cerebral blood flow. Ketamine produces a dissociative state associated with hallucinations on emergence and is the only intravenous induction agent with analgesic activity.

Inhalational Agents

Inhalational agents can be used for both induction and maintenance of general anesthesia as they produce unconsciousness, amnesia, analgesia and a small degree of muscle relaxation. Dosing and titration of these agents are affected by the solubility of the inhalational agent in blood, the patient's cardiac output and the minimum alveolar concentration of the anesthetic sufficient to prevent movement in 50% of patients. Agents with low solubility in blood have faster onset and recovery from the effect of the anesthetic. The inhalational anesthetics used in clinical practice include nitrous oxide, halothane, isoflourane, enflurane, and sevoflurane.

Nitrous oxide is a weak anesthetic and is administered in combination with other anesthetic gasses or during conscious sedation for its analgesic effect.

Halothane is one of the most commonly used anesthetic gasses but should be avoided in patients with increased intracranial pressure, cardiac disease, hypovolemia and liver disease secondary to its effects in causing cerebral vasodilation, nodal rhythm, ventricular irritability, hypotension and halothane hepatitis.

Isoflourane is only a mild cardiac depressant and does not induce ventricular irritability but has major side effects of increasing intracerebral blood flow and increasing the incidence of postpartum hemorrhage.

Enflurane is a more potent muscle relaxant, respiratory depressant and circulatory depressant than halothane and is metabolized to produce nephrotoxic fluorides.

Sevoflurane is the most effective agent for the cardiac patient with no significant effect on cardiac output, blood pressure, or peripheral vascular resistance. Sevoflourane is also useful in patients with increased intracerebral pressure.

Muscle Relaxants

Muscle relaxants are used to facilitate intubation in the induction phase and provide effective surgical exposure during the maintenance stage of the procedure (Table 4.2).

Succinylcholine is used mostly for tracheal intubation at induction due to its quick onset of action and rapid spontaneous recovery. Succinylcholine has many well known side effects:

1. Increased intracranial pressure
2. Increased intragastric pressure (predisposing to aspiration)
3. Sinus bradycardia and asystole from repetitive doses
4. Increase in serum potassium (avoided in patients with burns and spinal cord injury >24 hours out)

Nondepolarizing muscle agents often cause hypotension and bradycardia and are used in conjunction with other agents to support cardiac function. Reversal of muscle relaxants can be produced by either through the use of pyridostigmine, edrophonium or a combination of neostygmine and atropine.

Table 4.2. Characteristics of muscle relaxants

Muscle Relaxant	Dose (mg/kg)	Duration	Comments
Depolarizing			
Succinylcholine	1	5 min	Hyperkalemia
Nondepolarizing			
Mivacurium	0.2	10 min	Useful for intubation with quick onset and short duration
Atracurium	0.5	1 hour	Hoffman elimination without hepatic elimination
Vecuronium	0.1	1 hour	Rapid onset with little cardiac effects
Pancuronium	0.1	>1 hour	

Analgesics

Analgesia, although partially created through use of inhalational agents and ketamine, is primarily the domain of the opioids (Table 4.3).

Meperidine is the most effective agent in stopping postoperative shivering. Meperidine is metabolized to normeperidine that has significant CNS toxicity with prolonged administration or in patients with renal insufficiency.

Sufentanyl has a similar potency to fentanyl with shorter recovery times and greater cardiac and respiratory depressant effects. Alfentanil is only one fifth to one third as potent as fentanyl but has faster times to awakening, orientation and ambulation. The central nervous system effects and respiratory depression caused by opioids can be reversed using naloxone which also acts to partially reverse the effects of benzodiazepines and barbiturates.

Malignant Hyperthermia

When succinylcholine and inhalational agents are used, the patient must be observed for the development of malignant hyperthermia. Malignant hyperthermia is an inherited autosomal dominant defect in calcium sequestration in muscles leading to muscle contracture, increased oxygen consumption, increased lactate production and increased heat production after certain types of anesthesia. This is evident clinically through patient obtundation, muscular rigidity and hyperthermia. Management is with the use of intravenous dantrolene.

Table 4.3. Analgesic agents

Drug	Dose (mg/kg)	Class	Effect	Comments
Morphine	1	Opioid	Analgesia, sedation	Respiratory failure, broncho spasm, urinary retention, ileus
Meperidine	10	Synthetic opioid	Analgesia, sedation	Atropine-like effect, myocardial depression
Fentanyl	0.1	Synthetic opioid	Analgesia, sedation	Rapid metabolism, minimal cardiac and respiratory effects. In large doses it can produce trunkal rigidity.

References

1. Cousins M, Bridenbaugh P. Neural Blockade in Clinical Anesthesia and Management of Pain. Philadelphia: Lippincott-Raven, 1998.
2. Miller R et al, eds. Anesthesia. 5th ed. Philadelphia: Churchill-Livingstone, 2000.
3. Weinerkronish J, Gropper M. Consious Sedation. Philadelphia: Hurley & Belfus Inc., 2001.
4. Yaksh T et al, eds. Anesthesia Biologic foundations. Philadelphia: Lippincott-Raven, 1998.

4 Regional and Epidural Anesthesia

Regional Anesthesia

Regional anesthesia can be produced through local infiltration, nerve blocks or spinal/epidural anesthesia. Regional anesthesia can be induced by drugs that fall into two classes: the aminoesters and the aminoamides. Anesthetics with a pH close to neutral have a rapid onset of action and those with intense protein binding have an increased potency. The duration of action is determined mostly by local absorption which can be attenuated by mixing the agent with epinephrine.

Cocaine is rarely used to anesthetize the mucous membranes of the upper airway by topical application causing vasoconstriction and blocking uptake of norepinephrine at sympathetic nerve endings. Procaine is used for local infiltration of field blocks lasting from 30 minutes to one hour in duration.

Lidocaine is the most popular local anesthetic and can be used for local infiltration, nerve blocks, and spinal anesthesia. Bupivicaine is a potent amide with a long duration of action.

Side Effects and Toxicity

Anesthetics used for regional pain control exert systemic toxicity from either injection into the venous system or from extensive local infiltration (Table 4.4). Central nervous system effects progress from excitation with tinnitus and convulsions to eventual coma and respiratory arrest. Toxic concentrations of these anesthetics can produce depression of myocardial excitability, conductivity and contractility. Cardiotoxicity of these agents is increased by hypoxemia, hypercapnia, and acidosis. Allergic reactions can also be produced by local anesthetics including dermatitis, bronchospasm and anaphylaxis.

Table 4.4. Local anesthetics

Drug	Class	Duration of Action (Hours)
Lidocaine	Aminoamides	1
Bupivicaine	Aminoamides	4-12
Chlorprocaine	Aminoesters	0.5-1
Tetracaine	Aminoesters	0.5-1

Local Infiltration and Nerve Blocks

Local anesthetics with prolonged periods of analgesic activity such as bupivicaine are being used to provide postoperative pain control in outpatient surgery. Nerve blocks can be used to provide anesthesia for more extensive surgical procedures to the extremities.

Procedures

Digital nerve blocks can be done at the base of the fingers and toes by injecting proximal to the two dorsal and palmar/plantar nerves supplying each digit.

Brachial plexus block done by an interscalene, supraclavicular or axillary approach can provide anesthesia for the shoulder and hand.

Lower extremity blocks can be performed through injection of individual nerves including block of the femoral, sciatic, obturator and femoral lateral cutaneous nerves or through a lumbar plexus block involving injection of anesthetic into the psoas muscle to block all of these nerves simultaneously.

Ankle block involves individual block of the five nerve branches innervating the foot.

Bier block. Intravenous regional anesthesia can be provided through a Bier block where local anesthetics are injected into the limb's circulation isolated from systemic perfusion by a tournequit. Care must be taken to avoid injection of anesthesia into the systemic circulation or directly into the nerve causing a neuropathy.

Spinal Anesthesia

Spinal anesthesia is advantageous in patients requiring operations to the perineum and extremities in that it obviates the risk of aspiration and a difficult airway, allows for early identification of cardiac ischemic events and angina, and allows for detection of complications of urologic surgery such as bladder perforation and systemic absorption or irrigation solutions.

Despite these advantages spinal anesthesia has not shown a uniform reduction in cardiac ischemic events during surgery. Contraindications to spinal anesthesia include major coagulopathy and unstable neurologic disease.

Procedure

Spinal anesthesia is produced by injecting anesthetics into the subarachnoid space at the L3-L4 level and the spinal level of anesthesia is determined by the dose of the drug administered. Other factors that can determine the level and number of spinal segments anesthetized include pregnancy, increased intra-abdominal pressure, positioning of the head of the patient, injection at higher spinal levels, rapid injection of the anesthetic and barbotage (a technique that promotes mixing of the CSF with the anesthetic).

Complications

Complications from spinal anesthesia arise from the loss of sympathetic tone and leakage of cerebrospinal fluid. The most immediate complication is a drop in blood pressure without a compensatory increase in heart rate and is seen most commonly in geriatric and pregnant patients. Hypotension is easily remedied by fluid bolus or ephedrine. Focal neurologic defects following spinal anesthesia have also been reported but are minor and transient. Rarely patients can develop adhesive arachnoiditis with nonspecific inflammation of the meninges and spinal cord resulting in paraplegia.

Spinal headache is a second complication produced by intracranial hypotension. Treatment of spinal headache includes fluid boluses, treatment with caffeine, or an injection of blood into the epidural space at the level of lumbar puncture known as an autologous blood patch.

Epidural Anesthesia

Background

Epidural anesthesia is used for analgesia in operations in the abdomen, perineum and extremities as well as being used for postoperative pain control. Epidural anesthesia used during the course of general anesthesia and recovery has been found to reduce postoperative complications compared with the use of intravenous opioids alone for analgesia.

Epidural analgesia shows a greater preservation of preoperative pulmonary function, decreases myocardial ischemia, decreases the incidence of thromboembolic events and hastens the recovery of bowel function when compared with intravenous opioids. Epidural analgesia also increases blood flow to the viscera seen in lower anastomotic leak rates for procedures using a combination of general and epidural anesthesia.

Procedure

Highly lipophillic opioids such as fentanyl are combined with local anesthetics such as bupivicaine in continuous infusions to provide optimal analgesia with the least side effects. The epidural space extends from the foramen magnum to the sacrococcygeal membrane, and local anesthetics can be deposited in this space to block neural transmission diffusing intrathecally to block nerve roots. After catheter placement subarachnoid or intravenous cannulation is ruled out by injecting a test dose and monitoring for spinal anesthesia or tachycardia. The dose of anesthetic varies according to the level of anesthesia desired. Complications include arterial hypotension, transient backache, dural puncture, total spinal anesthesia, and epidural hematoma. Contraindications to epidural anesthesia include major coagulation defects, uncorrected hypovolemia, infection at the needle insertion site and patients with unstable neurologic disease.

Conscious Sedation

Conscious sedation is the practice of providing sedation, relaxation and analgesia to patients undergoing procedures as outpatients, in the emergency room or the intensive care unit without undergoing the hazards of tracheal intubation. Proper monitoring devices must be used to assure maintenance of the airway and cardiovascular support. The level of sedation and amnesia required is determined by the procedure and the agitation of the patient. As a rule the patient should be given analgesics and sedatives with short onset of action and quick recovery times. Versed is commonly used as a sedative because of its short onset of action and lack of residual sedative effect. Sufentanil and alfentanil are superior to morphine and demerol in providing analgesia for short surgical and diagnostic procedures. Agents such as propofol and ketamine are best limited to the ICU where patients can be closely monitored for perturbations in their respiratory status. Effective monitoring of vitals allows for titration of the analgesic and sedative effects of these agents to provide effective anesthesia for procedures and diagnostic tests of short duration.

References

1. Cousins M, Bridenbaugh P. Neural blockade in clinical anesthesia and management of pain. Philadelphia: Lippincott-Raven, 1998.
2. Miller R et al, eds. Anesthesia. 5th ed. Philadelphia: Churchill-Livingstone, 2000.
3. Weinerkronish J, Gropper M. Consious Sedation. Philadelphia: Hurley & Belfus Inc., 2001.
4. Yaksh T et al, eds. Anesthesia biologic foundations. Philadelphia: Lippincott-Raven, 1998.

4

CHAPTER 5

Statistics
Robert C. Kanard

Basic Medical Statistics

49% of all statistics are made up. —*Anonymous*

Background

Once data is paired down to its essentials, statistics continues to be needed to solidify and test conclusions. Untrustworthy data is no more than a compendium of interesting stories told by someone in a white coat.

Accuracy describes how far a data point is from the true value that is being measured. In archery terms accuracy describes how far away from the bull's-eye an arrow hits; each arrow is considered on its own. Accuracy correlates well with validity, which measures how far data deviates from the true value.

Precision describes how far a data point lies from the rest of the data points. It gives an idea of how reproducible a result is, and indicates the degree of random error. To use the archery metaphor again, precision describes how far an arrow hits the target away from the other arrows, regardless of where the arrows are in relation to the bulls eye. Like validity to accuracy, reliability correlates to precision.

Reliability is a measure of the reproducibility of a result: how many arrows can be shot into the same area on the target, regardless of their relation to the bull's-eye.

Mean, aka "average", is calculated by summing the value of all the data points and dividing by the number of data points. This gives a static picture of how a series of data has performed over a stretch of time. (Since it is algebraic in nature, it is by definition static; a dynamic look at how data changes over a set time period would require calculus.) For example, Mr. Smith's serum glucose levels over three days are 156, 240, 68,160,156,240,110,378,143,240,122, and 156. The mean of his serum glucose over these three days is calculated by adding all the values and dividing by 12 (four values each day for three days): 2169/12 = 180.75.

Median represents the middle data point in a series arranged sequentially (ascending or descending); and if the data set has an even number of observations, the two values in the middle define the median. The advantage of this is the median is not affected by the extremes (i.e., the 68 and 378 are effectively thrown out). In the example of Mr. Smith's glucose, arranging the values least to greatest: 68, 110, 122, 143, 156, 156, 156, 160, 240, 240, 240, 378, demonstrates the two middle values happen to be the same: 156.

Mode, simply put, is the most redundant value in a data set. In the example used thus far, 156 and 240 occur most commonly, therefore they are the modes. This seems rather inconsequential given the current example; however its significance is demonstrated by considering a disease such as adrenocortical carcinoma. This tumor occurs mainly during two age groups: children under five years old and adults between the ages of forty and fifty years old. The two groups of numbers occurring

Current Concepts in General Surgery: A Resident Review, edited by William R. Wrightson. ©2006 Landes Bioscience.

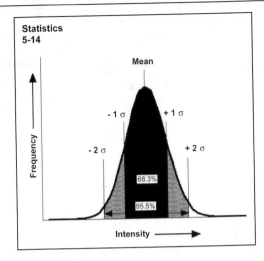

Figure 5.1. Normal bell shaped curve.

most often in a listing of ages of people with adrenocortical carcinoma are under five and forty to fifty year olds. This distribution is called bimodal.

Standard deviation is a term to describe how far certain data points lie from the center, or true, values; in other words, how accurate a value is. If a value is too far from the center (greater than two standard deviations), it may be discounted as inaccurate or incorrect. This is best conceived of by remembering the familiar picture of the bell-shaped (Gaussian) curve (Fig. 5.1).

This curve graphically describes a set of data points, the mean is the peak of the bell, but the median and mode are also represented, and in fact, the mean, median, and mode are all equal in the curve. The standard deviation describes where in relation to the mean a data point is. The significance of this is that 68% of a group of data points lie within one standard deviation of the mean. Furthermore 95% of the group will fall within two standard deviations of the mean, and 99% lie within 2.5 standard deviations of the mean. When reading a study, if the data looks too good to be true, and most of the supporting data lie within four standard deviations, it probably is too good to be true. This can also be used for an individual patient such as Mr. Smith, where the further a patient's data point (a serum glucose of 378) is from the center, the more likely it is to be an abnormal value and not a normal variant.

Incidence is the number of new cases reported in the total population. Recently the incidence of HIV infection has decreased (considered over all populations) as testing, screening blood products, and education of at-risk populations has improved.

Prevalence is the number of existing data points (patients with a disease) in a population. Although the incidence of HIV infection has declined over the past several years, the prevalence has increased. How is this possible? People are still being diagnosed with the disease every year, just at a slower rate, and those with the disease now live longer with the advent of protease inhibitors. The end result of these two forces are the increase in the total number of people with HIV: an increase in the prevalence.

Table 5.1. Disease state

Test Result	Afflicted	Not Afflicted
Positive	A	B
Negative	C	D

Frequency is the number of afflicted people out of the at-risk population only. This effectively limits the population under consideration in incidence and prevalence to those at risk.

Relative risk shows a comparison of the people with a disease, who have, versus have not, been exposed to a certain risk (e.g., pathogen, carcinogen). The population under scrutiny is that group with the disease, and relative risk tries to answer the question, "is exposure to X responsible for Y disease?" A relative risk greater than one means there is more disease in the exposed population, therefore a positive association exists between the exposure and the disease (children exposed in utero to thalidomide are born with a higher rate of phocomelia than those children not exposed in utero to thalidomide). A relative risk equal to one shows that there is no relation between exposure and disease (children exposed in utero to caffeine are not born with a higher rate of phocomelia than those children not exposed in utero to caffeine). A relative risk less than one means there is less disease in the exposed population than in the unexposed population. It does NOT mean that exposure conveys a protection or immunity to the disease; that is a separate issue. Relative risk can be calculated only for prospective studies.

Attributable risk mathematically is the incidence rate in people NOT exposed subtracted from the incidence rate in the exposed population. If two groups of 1000 babies each (one exposed to thalidomide in utero and the other not) are studied, and 100 babies out of the former group are found to have phocomelia, while 3 out of the latter have the deformity, the attributable risk is 97. That is, 3 diseased in the unexposed group subtracted from the 100 in the exposed group, which can be stated as the risk of phocomelia attributed to thalidomide exposure in utero is 97 out of 1000. This can be useful in predicting what would happen in a population if the exposure risk was removed.

Odds ratio is an estimate of relative risk calculated from retrospective studies (discussed below). The odds ratio is like looking at relative risk through the looking glass. Now that a group of people with a particular disease has been identified, can the odds that those affected people were exposed to something be compared with the odds that a group of control subjects (healthy population) was exposed? The answer is yes, and it can be expressed as the odds ratio. This is useful when the incidence of a disease is unknown, but the relationship between exposure to an agent and a disease is still being examined.

Knowing this vocabulary lays the basis for using the tools of statistics. Some of the studies have already been referenced, but discussion about these tests and studies begins with that famous "2 x 2 box" (Table 5.1).

Sensitivity measures how well a test identifies people with a disease. Graphically speaking, sensitivity is calculated as A/A+C. In other words, sensitivity begins with the population of all those afflicted and measures the odds that someone with a positive test actually has the disease. Of course, a higher sensitivity is preferable to lessen the chances of false negatives, which represent afflicted people, who test negative. A second issue relating to sensitivity is positive predictive value, which measures the accuracy of a test. The positive predictive value measures the odds that someone with a positive test actually has the disease. Graphically this can be represented by A/A+B. Because the starting population for sensitivity is that group of afflicted people, and the positive predictive value is a measure of the accuracy of a test, the positive predictive value is highest when the disease prevalence is high. More people with a disease means more people in the sample population (for sensitivity), and thus measures (positive predictive value) regarding the population under scrutiny are stronger and more exact (variation is minimized by increasing the denominator).

Specificity measures how well a test identifies people without a disease. Graphically speaking, specificity is calculated as O/B+O. As opposed to sensitivity, specificity concerns itself with the healthy population and measures the odds that someone with a negative test actually does not have the disease. A higher specificity is desirable to decrease the number of false positives, which represent those testing positive, but who are actually not afflicted. Unfortunately, as often occurs, as the specificity increases, the sensitivity decreases. The ideal test has both a high specificity and sensitivity. Another term relating to specificity is negative predictive value, which measures the odds that a person with a negative test actually is healthy. Graphically speaking this can be written as O/C+O. Because the negative predictive value represents the ability of a test to identify healthy people, it is highest when the prevalence of a disease is low.

Types of Studies

Case series is the most basic form of study. It looks at a cluster of people with a disease. Often it is the initial report of a new disease. It may ask questions and propose hypotheses, but it cannot be used to test them or pretend to provide any hard data on risk of exposure or consequence of therapy.

Cross-sectional studies, the next step, study the possibility of linking risk factors with disease states. They study a group of people at one particular point in time; therefore these studies only look at the prevalence of a disease against the backdrop of an exposure. Since they are temporally static (one group at one point in time), they shed no information on the relationship between a risk factor and a disease in terms of which came first. These studies are useful for developing hypotheses, including those raised by the initial case reports.

Retrospective studies, also called case-control studies, compare afflicted individuals ("cases") to those unafflicted ("controls"). These two groups are identified, interviewed, and data (especially exposure to known, or sometimes unknown, risk factors) is retrieved from the information the subjects provide. From this the odds ratio can be calculated, which, as stated above, approximates the relative risk (the relationship between disease and exposure). The advantages to this kind of study are that few subjects are needed, which makes these relatively quick and inexpensive to perform. This format lends itself well to rare diseases and those with a long latency period. The main disadvantage to this type of study is the reliance of the data on the subjects' memories. This can lead to a significant recall bias, whereby memory causes the results to deviate from the truth. Another disadvantage to these types of studies is the inability to calculate disease incidence from the data, because the diseased group and the healthy group are preselected.

Prospective studies, also known as cohort studies, begin with a group of individuals ("cohort"), who may, or may not, share a common exposure, but it characterizes members of the cohort based on different features (e.g., habits, geography, income) The study then follows them over time and evaluates their health as some are exposed to certain risks and others are not. A good example is The Framingham Heart Study. Advantages to prospective studies include being able to examine a multitude of exposures and diseases and the relative risk can be calculated from this type of study. It is important to have a method to calculate the relative risk because ultimately that is the goal: what is the cause/effect relationship between an exposure and a disease?, an intervention and a change in a patient's status? Also, obviously, recall information depends on the relationship between a risk factor and a disease in terms of which came first. These studies are useful for developing hypotheses, including those raised by the initial case reports. Also, obviously, recall bias is not an issue since the production of data paces its collection. The main disadvantage to prospective studies is the beginning cohort (the very premise of the study) is a group of otherwise healthy individuals who presumably have the same rates of exposures and disease as the general population. Therefore it is exceedingly difficult to study rare diseases or disorders, and a large number of subjects are needed to adequately perform the study. This then leads to two more drawbacks of cohort studies: they are more expensive and may take years to collect the necessary data. The gold standard of testing for the evaluation of a new intervention is the randomized, double-blinded, placebo-controlled clinical treatment trial, a type of prospective study. In these, subjects are arbitrarily assigned to one of two groups (randomization). One cohort then receives the treatment, while the other receives the standard of care or at least an "inert" treatment (placebo-controlled). Because neither the examiner nor the subject knows who receives which, the design is termed double-blinded. These four qualifiers reduce bias by the greatest margin.

A **hypothesis**, in its most basic form, is a statement that postulates a difference exists between two or more groups. For example, a new cholesterol-lowering drug X is compared to an old drug Y. The hypothesis for this comparison could state (blandly), "treatment cohort A has a lower total serum cholesterol than cohort B, because treatment cohort A was treated with drug X, while cohort B was treated with drug Y." This draws a distinction between cohorts A and B. Once a hypothesis is created, it must be tested and withstand scrutiny. Studies are conducted, but the hypothesis must still be statistically proven to be true to be accepted. To do this, statistics employs a special tool to test the validity (effectively the sensitivity and specificity) of a hypothesis. This tool is simply the opposite of the hypothesis. If the hypothesis claims a difference between two groups, the null hypothesis claims that no significant difference exists between the groups. Statistical analysis proceeds to either accept or reject the null hypothesis. By rejecting the null hypothesis (the antithesis of the hypothesis), the hypothesis is assumed to be valid, and a significant difference is assumed to exist between the groups. To the researcher the null hypothesis has but one function: rejection.

Probability refers to the number of times a result would recur, if an experiment were repeated indefinitely. Probability is quantified as the p value, and quantifies how strong a connection exists between an outcome and an intervention (hypothesis), while the null hypothesis states that a result and an intervention are random occurrences. If the p value is equal to or less than 0.05, it is likely that the null hypothesis is rejected, and the hypothesis is true. Two types of "errors" can be made when considering a hypothesis by making erroneous assumptions from data or if data conceals a bias (a force that deviates data from the true values). A type I error is committed, when the p value is equal or less than 0.05 and the null hypothesis is rejected, despite it being true. Because if the null hypothesis is true (drug X has no effect on serum cholesterol levels; it was random chance that cohort A has lower serum cholesterol levels), then it must be accepted, and therefore the hypothesis is false. It is an error to reject the null hypothesis under these circumstances. Therefore a p value less than 0.05 is considered "statistically significant", assuming no error is made. Likewise it is an error to accept the null hypothesis even if it is false (type II error). It requires reconsideration of the data or the study design to determine that the hypothesis is in fact true despite the null hypothesis being accepted.

It is important to remember that between the different kinds of studies, boxes, and validity tests, the subjects under consideration remain medicine and human disease. If some information does not match clinical experiences or training, then it likely warrants further investigation. Also just because a hypothesis is found to be true and a particular intervention is statistically significant, that does not convey clinical significance.

References

1. Fadem B. Behavioral science. Philadelphia: Harwal Publishing, 1994.
2. Feibusch K, Breaden R, Bader C et al. Prescription for the boards: USMLE step 2. Philadelphia: Lippincott-Raven Publishers, 1998.
3. Hill A. A short textbook of medical statistics. Philadelphia: JB Lippincott Company, 1977.
4. Mack S. Elementary statistics. New York: Henry Holt and Company, Inc., 1960.
5. Weinberg G, Schumaker J. Statistics, an intuitive approach. Belmont, CA: Wadsworth Publishing Company, Inc., 1962.

CHAPTER 6

Wound Healing

Robert I. Oliver, Jr.

Physiology of Wound Healing

Background

The classic model for wound healing involves three distinct phases:

1. Inflammation
2. Proliferation/Fibroplasia
3. Contraction/Remodeling

Inflammation

The inflammatory phase begins with an injury or insult that stimulates an inflammatory response by exposure of subendothelial collagen to platelets, which activates the intrinsic and extrinsic clotting cascades and causes platelet degranulation, with release of cytokines and growth factors from platelet α-granules. These platelet-produced growth factors like platelet-derived growth factor (PDGF) and transforming growth factors alpha (TGF-α) and beta (TGF-β) are released locally in the wound and amplify the recruitment of PMNs, and more importantly macrophages. After a brief vasoconstrictive phase mediated by *thromboxane* and *serotonin* that aids in hemostasis, a vasodilatory reaction (in response to platelet *histamine* release and *prostaglandin* generation) follows with migration of leukocytes, neutrophils (PMN), and later, macrophages into the wound along the fibrin clot scaffolding.

PMNs are the predominant cell type early on and migrate into the wound through an increase in vascular permeability and track along chemotactic gradients compromised of cytokines, complement cascade products (C3a and C5a), and growth factors in a process called margination. Endothelial receptors called selectins help PMNs loosely adhere to the involved area before integrin receptors on the PMN allow firm interaction and allow subsequent migration between endothelial cells into the extracellular matrix (ECM).

PMNs are an important adjunct, but not a necessity, for wound healing. They serve to promote bacterial clearance and debride the wound, but provide no essential growth factors.

Monocytes migrate into the wound and undergo transformation into activated macrophages at close to 48 hours, and their subsequent production of inflamatory cytokines (IL-1 and TNF) and growth factors (specifically transformation growth factor-β [TGF-β] and PDGF) appears to be the most critical cell-driven event of this entire phase. Their absence is associated with failure to progress to normal fibroblast recruitment and function. This is in contrast to experiments which show wound healing to progress in the absence of PMNs and lymphocytes. Macrophages also serve a number of other important functions including nitric oxide synthesis, wound debridement, phagocytosis of bacteria, and stimulation of angiogenesis.

Current Concepts in General Surgery: A Resident Review, edited by William R. Wrightson.
©2006 Landes Bioscience.

Proliferation and Fibroplasia

This proliferative phase of wound healing starts 48-72 hours after the injury and as is characterized by fibroblast activation and infiltration, and subsequent synthesis of the extracellular matrix (ECM) and collagen. Fibroblast function is driven by platelet and macrophage-produced growth factors (most importantly TGF-β). The ECM is composed of a group of complex proteins called glycosaminoglycan (GAG) and hyaluronic acid, which with the addition of fibronectin, the primary adhesive protein of the ECM, serve as a scaffold for migrating cells and nutrient diffusion into the wound. Fibronectin also serves as a chemoattractant for macrophages as well as to assist in activating them locally.

Macrophages are also actively involved in this phase by secreting growth factors that promote angiogenesis by inducing proliferation of endothelial cells from adjacent vessels. These endothelial cells establish a new capillary network in the wound until a rising oxygen tension serves as a negative feedback to the process. Angiogenesis begins at the periphery of the wound from capillary buds that branch off of neighboring venules and proceeds to arborize into a fine vascular network across the wound.

In a normally healing wound, by the end of the first week, few remaining PMNs and macrophages should be present in the wound. The inflammatory phase may become persistent however, due to the presence of necrotic material, foreign bodies, or bacterial colonization and delay the normal progression of healing for some time until they are removed by the host defenses or by medical/surgical care of the wound.

Epithelialization/Contraction

The processes of epithelialization and contraction of the wound also occur during the proliferative phase. Epithelialization is the process where neighboring basal epithelial cells proliferate and begin to move into the wound from the edges to establish a barrier to fluid-loss and infection as they layer out across a wound surface. As the leading edges of these migrating cells contact one another, they undergo contact inhibition to arrest further migration and then proceed to reestablish a true multilayered epidermal layer. Epithelial buds also form from intact epithelial appendages in the middle of the wound, such as hair follicles and sweat glands. Well-approximated surgical wounds reepithelialize as rapidly as 24-48 hours and heal by the process called primary intention. Contraction on the other hand is the process where wound edges are mechanically approximated by contractile forces generated by special fibroblast called myofibroblasts, which have rudimentary actin-myosin machinery. The process where wounds undergo healing mainly by contraction is called secondary intention and occurs normally at a rate of just under 1 mm/day. Evidence of contraction can be seen clinically at approximately one week post-injury. In approximated surgical wounds, contraction contributes little to the healing process. Coverage of wounds with split-thickness or full-thickness skin grafts and delayed primary closure of a wound are techniques that are able to minimize excessive contraction of wounds, which can result in scar contractures if joints are involved.

Remodeling

The remodeling phase begins after collagen production/degradation has equilibrated (though both process continue at accelerated rates), usually some three weeks after the inflammatory phase began. Collagen synthesis reaches its maximum rate at close to one week while levels in the wound elevate for up to three to four weeks before reaching equilibrium. The net synthesis of collagen however, persists for up to 4-5 weeks. Characteristic of the remodeling phase is the reorientation and cross-linking of the collagen fibers into more organized patterns and the replacement of immature type III collagen with type I collagen fibers. Cross-linking of collagen is what increases the bursting strength of the scar. The strength of wounds increases linearly by approximately 3% at one week, 50% strength is achieved after six weeks, and a maximum strength of up to 70-80% of prewounded tissue is achieved at three months.

Collagen

Collagen is the dominant fiber present in connective tissue and is largely responsible for the tensile strength of wounds. In its mature form it has a complex three-dimensional structure of three peptide chains each twisted in a right-handed helix with the complex of the three chains then arranged in a left-handed superhelical formation. The collagen molecule is secreted by activated fibroblasts and is characterized by a repeating sequence of glycine (gly)-X-Y with X usually being proline, and Y usually being hydroxyproline or hydroxylysine. The hydroxylation of lysine is important for the covalent cross-linking of fibers and is facilitated by the enzyme lysyl oxidase. Ehlers-Danlos syndrome is a genetic defect involving inadequate production of lysyl oxidase (an enzyme involved in the cross-linking process), which produces defective or weak collagen and is associated with poor wound healing. Failure to hydroxylate proline affects collagen transport out of the cells. Vitamin C is a cofactor necessary for these hydroxylations of collagen residues, and its absence produces the clinical condition of scurvy with its associated deficient, defective, and weak collagen. Hypoxia or corticosteroid administration also retard hydroxylation and produce similarly defective collagen. A series of post-translational steps produces the procollagen molecule, which is the secretory form, with its characteristic N and C terminal propeptides. Once in the ECM, specific N and C terminal proteinases remove the terminal peptides and allow the organization of collagen into fibrils to begin.

Once relieved of its terminal proteins, collagen assumes a complex structure of three polypeptide chains in a helical formation that are covalently-bonded to each other to form a tropocollagen molecule. Tropocollagen molecules subsequently aggregate to form collagen filaments, fibrils, and eventually the macromolecular collagen fibers.

Close to 20 different collagen molecules have been identified and characterized. Type I collagen makes up over 90% of the collagen in the body and is the dominant type in mature wounds. Type III collagen is also a key in early wound healing, composing up to a third of all wound collagen during the granulation tissue of the fibroblastic phase, before being replaced during the remodeling phase to restore the normal 4:1 ratio between types I and III collagen that exist in normal skin and mature wounds.

Growth Factors in Wound Healing

Growth factors are hormone-like proteins that affect metabolism, growth, and differentiation of cells during all three phases of wound healing. Multiple cell lines and function secrete them by paracrine, endocrine, intercrine, and autocrine mechanisms to alternately stimulate or inhibit the processes of wound healing by interacting with unique cell-surface receptors. Signal transduction is accomplished by various

Table 6.1. Major growth factors in wound healing

Factor	Source	Function
PDGF	Platelets, macophages, fibroblasts, endothelial cells	Fibroblast and smooth muscle chemotaxis proliferation, collagen synthesis
TGF-β	Macrophages, platelets, fibroblasts	Potent collagen synthesis inducer, stimulates wound contraction, stimulates angiogenesis
TGF-α	Macrophages, platelets, keratinocytes	Stimulates proliferation of epithelial, endothelial, and fibroblast cells. A very potent angiogenesis stimulator.
EGF	Macrophages, platelets, keratinocytes	Similar to TGF-α
IL-1	Macrophages, PMNs	Inflamatory cell stimulation chemotaxis, collagenase synthesis
TNF	Macrophages, lymphocytes	Fibroblast stimulation, collagen synthesis, angiogenesis
FGF	Macrophages, endothelial cells	Angiogenesis, collagen synthesis, wound contraction, ECM synthesis

6

second messenger systems including tyrosine kinases, G-proteins, DAG/IP-3, and protein kinase C which all serve to alter gene expression and stimulate protein synthesis. An absolute or relative deficiency of growth factors in a wound may contribute to wound failure and production of the chronic wound environment. A summary of the major identified growth factors is included in the Figure 6.1 below (Table 6.1).

Chronic Wounds

The chronic wound is one that fails to proceed thru the normal progression of wound healing in a timely fashion. Characteristic of chronic wound healing is prolific granulation tissue deposition with significant fibrosis/scar formation of the tissue. Granulation tissue is a rich collection of capillary buds interspaced with multiple cell lines including fibroblasts and inflamatory cells. As healing progresses granulation tissue normally undergoes apoptosis to an avascular and acellular collagen matrix. Persistence of granulation tissues results in hypercellular tissue with eventual hypertrophic scar formation.

A large number of identifiable factors predictably will impair, delay, or prevent satisfactory resolution of a wound. These factors are summarized in Table 6.2A and Table 6.2B.

A special group of problem wounds worth noting include venous stasis wounds and pressure sores. Venous ulcers result from progressive lower extremity venous valvular failure with resultant edema formation, microcirculatory occlusion, and perivascular fibrin cuffing. Eventually breakdown of the overlying skin results from the hypoxic state and frequently troublesome relapsing wounds develop. In addition, stasis ulcers have a relative deficiency of growth factors due to decreased breakdown by proteinases and by the sequestering of growth factors in the perivascular fibrin cuffs. Surgical treatment by skin grafting the wound or ligation of subfascial

Table 6.2A. Factors affecting wound healing

Radiation	Radiation produces both endothelial cell and fibroblast cell injury, affecting both perfusions of the tissue as well as affecting collagen and ECM secretion. Decreased infiltration by inflamatory cells also lowers available cytokine amounts, and impairs all phases of healing to some degree.
Malnutrition	Low protein states affect available amino acids available for collagen synthesis. Fatty acid and carbohydrate deficiencies retard wound healing in part by stimulating protein breakdown to supply calories.
Aging	Characterized by a linear decrease across the board in the wound healing processes (contraction, epithelialization, cell migration, collagen/ECM synthesis)
Infection	Heavily colonized wounds demonstrate prolonged inflammatory phases and decreased fibroblast proliferation and function
Hypoxia	A multifactoral process contributed to by smoking, diabetes, peripheral vascular disease, edema, and radiation. Hypoxic tissue is prone to infection and heals poorly. The increased intercapillary distance in wounds requires higher tissue PO_2 to drive oxygen into healing areas. Oxygen is necessary for epitheliazation, collagen synthesis, and bacterial destruction by oxygen generated free radicals.
Steroids	Affects DNA synthesis during the inflamatory phase by downregulating DNA synthesis of inflamatory cytokines. This effect is blunted somewhat by dietary supplementation with Vitamin A (10-25K international units Q day)
Diabetes	Affects the inflammatory process as well as damages regional microcirculation
Smoking	Lowers tissue PO2 by its vasoconstrictive properties and contributes to atherosclerosis. Smokers also have elevated carboxyhemoglobin levels.
NSAIDS	Decrease collagen formation
Chemotherapy	Decrease collagen synthesis and fibroblast proliferation

perforating veins (the Linton procedure) has been for the most part unsuccessful, although they do have some proponents. Sensible and cost-effective strategies for treating stasis ulcers include fitted or off-the-shelf compression garments and leg elevation with the hope of eventual spontaneous secondary healing. A useful technique for assisting healing in noninfected stasis ulcers is the weekly application of Unna-boot™ wraps, a commercially available semi-rigid gauze impregnated with zinc oxide paste, which serves to assist with the calf-muscle pump for venous return as well as to allow prolonged contact with the zinc paste, which may serve to

Table 6.2B. Vitamin and mineral deficiency

Vitamin C	Cofactor for lysyl hydroxylase (collagen hydroxylation)
Vitamin A	Cofactor for epitheliazation and collagen synthesis
Vitamin E	Strong anti-oxidant effects
Zinc	Cofactor for DNA synthesis
Copper	Cofactor for lysyl oxidase (collagen cross-linking)
Vitamin B$_6$	Cofactor for collagen cross-linking

promote healing. Oral pentoxyifylline is an agent that has shown promise as an adjunct to compression treatment, facilitating healing by some as yet unknown mechanism.

A number of very expensive topical growth factors (notably PDGF) and wound dressings have been marketed over the years with claims of facilitating closure of stasis ulcers, but to date these have been both disappointing clinically and economically unfeasible, and in general do not play a role in how most of these wounds are treated. In the future, as the etiology and biochemistry of these wounds is better understood, we can expect more successful chemical adjuncts for routine clinical use that demonstrate in vivo efficacy.

Pressure ulcers are a group of wounds that frequently occur in a wide variety of patient populations including the elderly, debilitated, paralyzed, or nonambulatory groups. The common etiology is sustained pressure over a bony prominence resulting in microcirculatory compromise. This can occur and produce tissue necrosis with sustained pressures as low as 25 mm Hg for durations as short as two hours. Frequent sites for pressure ulcers include the sacral/ischial/trochanteric region, posterior scalp, posterior heels, metatarsal heads in diabetics, and on extremities with casts on. The most important strategy for treating pressure sores is prevention. Frequent turning and repositioning of bedridden patients is mandatory and serves to eliminate the majority of avoidable pressure sores. Treatment of existing wounds also involves avoiding pressure to the area, as well as local wound care with dressing changes and sharp debridement of devitalized tissues.

In dealing with a presumed venous stasis ulcer, one should not overlook the possibility that the etiology is ischemic or some combination of arterial insufficiency and venous stasis. A pulse exam should be done just as with peripheral vascular disease (PVD) and vascular noninvasive studies (i.e., ABI's with segmental limb pressures) and possibly arteriograms should be obtained in those without palpable distal pulses who are surgical candidates. The use of transcutaneous oxygen probes has also been reported to be of use in assessing these wounds, with a TCPO$_2$ of less than 30 predicting failure of spontaneous healing due to depressed fibroblast function. Nonhealing wounds in the face of PVD merit referral to a vascular surgeon for a potential revascularization.

Wound Dressings

An important technique of local wound care involves the use of saline-moistened dressing changes. As the moist gauze is applied to the wound and proceeds to dry out, necrotic material and debris in the wound adhere to it and are removed as the dressing is changed (i.e., "wet to dry" dressings). Frequent dressing changes (3 or 4 times daily) can serve to mechanically debride heavily colonized wounds but should not replace sharp debridement of grossly nonviable tissue. A variety of antimicrobial

Table 6.3. *Topical antimicrobials*

Agent	Comment	Notable Side-Effects
Bacitracin	Gram positive coverage only	
Betadine solution	Broad spectrum but short acting due to rapid dilution in the wound. Colloidal or gel preparations are signifigantly longer acting.	Toxic to fibroblasts at higher concentrations. Contraindicated in patients with iodine allergies.
0.025% Sodium Hypocholorite (Dakin's Solution)	Bacteriacidal to all fungi and bacterial species. The single best agent for lowering bacterial counts.	
Silver Sulfadiazine (Silvadene™)	Broad spectrum. Penetrates eschar poorly.	Associated with transient neutropenia. Contraindicated in patients with sulfur or silver allergies.
0.5% Silver Nitrate	Broad-spectrum bacteriostatic agent with the exception of *Klebsiella* and *Enterobacter* species. Penetrates eschar poorly. Painless on application.	Associated with electrolyte abnormalities (hyponatremia and hypochloremia) and rarely methemogolbinemia.
Mafenide (Sulfamylon)	Broad spectrum except for fungus and MRSA. Very good eschar penetration. Painful during application.	Can cause a metabolic acidosis (due to carbonic anhydrase inhibition), is painful on application, and lacks antifungal coverage.
Polymyxin	Gram negative coverage only.	

solutions can be used instead of saline, each of which has their own advantages and spectrum of activity. One should also familiarize oneself with some of the common side effects of these agents in order to recognize potential life-threatening complications (see Table 6.3).

Care must be taken when using moistened gauze dressing changes to avoid either macerating or drying out the wound excessively. A common strategy to prevent irreversibly desiccating the tissue is to change the dressing before it dries out completely (i.e., "wet to moist" dressings). To avoid maceration, the wet dressing should not be over moistened and have minimal contact with surrounding normal tissue by securing it in a fashion that it does not shift in and out of the wound. A useful strategy in wounds not requiring further debridement is to discontinue dressing changes and place Silvadene™ cream on the wound once or twice daily to both continue antimicrobial effects and to perpetuate a moist environment.

The goal of a wound dressing on colonized wounds should be to both protect the wound from the environment and to assist in lowering the bacterial count of the wound. Surveillance of bacterial growth in the wound is critical before performing any type of wound closure, be it by primary closure, skin grafting, occlusive dressing, or flap coverage. Simple swab cultures of the wound are a cost-effective way of determining a rough estimate of bacterial colonization, but quantitative tissue culture from wound biopsy is the gold-standard prior to performing closure or coverage planning. Bacterial counts greater than 10^5 colony-forming units (CFU) have been reliably shown to predict failure of grafts, flaps, or primary closure. Special attention should be paid to cultures that reveal beta-hemolytic Streptococcus, which can cause wound failure with as few as 100 CFU.

Occlusive dressings provide a moist environment favorable to more rapid epitheliazation. These types of dressings require close surveillance for development of infection under them and are more useful for fresh wounds that have had little chance to develop bacterial colonization. Suspicion of infection requires removing the occlusive dressing and using dressing changes or topical antimicrobial therapy. Occlusive dressings can be particularly useful for small cuts and scrapes, as well as for superficial burns and as a dressing on donor sites from skin grafts.

A recent advance in wound dressing is widespread adoption of the vacuum-assisted closure device (VAC™). This device employs an open-cell foam sponge placed in the wound that's covered with an occlusive dressing and then put to a powerful vacuum pump. When a proper seal is created, a form-fitting dressing is created. The VAC works by evacuating excess fluid and by facilitating removal of both bacteria and metalloproteinases (a group of enzymes that degrade growth factors) in the wound. The subatmospheric pressures generated also serves to physically facilitate contracture of the wound. Edema is deleterious because it serves as both a diffusion barrier to nutrients and as a diluent of locally produced antibacterial sebaceous gland secretions.

Advantages of the VAC wound dressing include not only faster epithelialization and wound contraction, but they also simplify care by requiring less frequent dressing changes (Q2-3 days) in clean wounds. VAC's have been used on many complex wounds including sacral decubs, perineal wounds, open abdominal wounds, and as a bolster for skin grafts. VAC dressings have also been applied to colonized wounds and have been effective clinically in decreasing bacterial loads. If used in this way however, the sponge may need to be replaced as frequent as twice-daily to survey for underlying infection and assist in debridement. At present, prudent recommendations for VAC dressing application should probably be limited to clean or lightly-colonized wounds.

Nutritional Support

An important and frequently overlooked source of wound failure is malnutrition. A thorough assessment by a clinical dietician should be obtained for any patient where nutritional status is in question. General recommendations for caloric needs start with approximately 25-35 kcal/kg/day of nonprotein calories (fats + carbohydrates) and 1-1.5 gm/kg/day of protein. Other than clues on physical exam, a history of alcohol abuse and serum albumin < 3.0 may suggest chronic malnutrition. An important and convenient serial test to follow for improvement is the protein, prealbumin, which has a half-life of just less than one week, as opposed to several weeks for albumin. An increasing prealbumin, checked weekly, is highly suggestive a patient in positive nitrogen balance.

A recent advance in nutritional support for wound healing involves supplementing the patient with the anabolic steroid, Oxandrin™ (oxandrolone). Taken as an oral supplement (10 mg twice daily), Oxandrin has shown exciting potential in the treatment of large burns by blunting the protein wasting associated with the hypermetabolic state post-burn. Clinically faster reepitheliazation of skin graft donor sites and the maintenance of lean body mass have reflected this. Results have also been reported in assisting closure of some chronic wounds.

A number of vitamin, electrolyte, and trace element deficiencies may impair healing. In patients with a normal, well-balanced diet, it will be rare that significant nutritional deficiencies will exist. However, in the hospitalized patient, a number of clinical deficiencies of these substrates may present. Supplementation with an oral or IV multivitamin is a cheap method of treating or prophylaxing against occult deficiencies. Additional dosing of supplemental zinc sulfate and vitamins A, C, and E are advocated by some for wound healing in larger or chronic wounds. Excess dosing of fat-soluble vitamins (A,D,E, and K) can lead to systemic toxicity, however.

References

1. Steed DL. The Role of growth factors in wound healing. Surg Clin North Am 1997; 77(3):575-586.
2. Stadelmann WK, Digenis AG, Tobin GR. Impediments to wound healing. Am J Surg 1998; 176(Suppl 2A):39S-47S.
3. Eaglstein WH, Falanga V. Chronic Wounds. Surg Clin North Am 1997; 77(3):689-700.
4. Clark RA. Wound Repair: Overview and General Considerations. The Molecular and Cellular Biology of Wound Repair. 2nd ed. New York: Plenum Press, 1996.
5. Brew EC, Mitchel MB, Harken AH. Fibroblast growth factors in operative wound healing. J Am Coll Surg 1995; 180:499.

CHAPTER 7

Critical Care

William R. Wrightson

Sepsis and Management

Background

Sepsis is a major cause of morbidity and mortality in the US with > 750,000 cases of severe sepsis reported annually. In the US > 500 patients die of sepsis daily. The mortality from sepsis ranges between 28% and 50%. Sepsis is defined as a series of syndromes from SIRS to severe sepsis.

SIRS

Systemic inflammatory response syndrome (SIRS) is defined as a clinical response arising from a nonspecific insult including ≥ 2 of the following:

- Temperature ≥ 38 C or ≤ 36 C.
- HR ≥ 90 beats/min
- Respirations ≥ 20/min
- WBC ≥ 12000/mm^2 or ≤ 4000/mm^2 or ≥ 10% immature neutrophils

Sepsis

Includes the above criteria with the addition of a presumed or identified source of infection.

Severe Sepsis

Severe sepsis included ≥ one sign of organ dysfunction (Fig. 7.1).

Etiology

The development of sepsis in multifactoral but is generally the result of activation of the systemic inflammatory response. This can be due a variety of sources including pancreatitis, trauma, critical illness, organ failure, and pneumonia.

Pathophysiology

The basis for sepsis is the activation of inflammatory mediatiors. Eary response includes the activation and release of TNF, IL-1 and IL-6. This results in the activation of platelets and the coagulation cascade. The net result is an increase in inflammation, coagulation and decreased fibrinolysis. This leads to endothelial dysfunction and microvascular thrombosis.

Role of Protein C in Severe Sepsis (Xigris® and Prowess Trial)

Activated protein C is an endogenous modulator with antithrombotic, anti-inflammatory and profibrinolytic properties. Underexpression of activated protein C plays a role in the progression of sepsis. Drotrecogin (Xigris®) is recombinate

Current Concepts in General Surgery: A Resident Review, edited by William R. Wrightson.
©2006 Landes Bioscience.

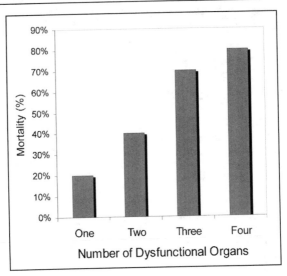

Figure 7.1. Mortality is directly associated with the number organ failures.

activated protein C that reduces inflammation, coagulation and increased fibrinolysis. The PROWESS trial evaluated the use of drotrecogin alfa in severe sepsis and found an improved survival at 28 days. Mortality was 24.7% in drotrecogin alfa treatment groups versus 30.8% in control groups. Patients at the highest rick (APACHE ≥ 25) had the greatest benefit with a 13% reduction in mortality. Those patients with APACHE ≤ 20 had the lease benefit. Bleeding complications with the use of drotrecogin alfa are 17% versus 8% in placebo groups.

Current dosing for Xigris is 24 μg/kg/hour over 4 days. In a recent symposium, patients with significant improvement the infusion was stopped early without any significant loss of benefit. This drug is expensive with a cost of infusion at $6600.00. The end result is 18 patients must be treated to save one life with Xigris.

Other Research and Treatment Strategies

Several studies have further evaluated strategies against the effects of endotoxin. These include oral feeding with glutamine in an endotoxin animal model, blockade of the testosterone receptor, and blocking calcium channels with diltiazem or verapamil.

A pilot clinical project evaluating the effects of methylene blue in human septic shock has been completed and appears to counteract myocardial depression, and maintain oxygen transport, through methylene blue's effect on inhibiting nitric oxide.

Several other studies reported this year have shown elevated levels of heat shock proteins in association with survival in sepsis, and increased activated leukocyte microparticles in sepsis. The relationship of nitric oxide to p38 MAP kinase activation and the effects of secretory IgA on maintaining the intact intestinal barrier after ischemia and reperfusion and bacterial challenge have also been further clarified.

Two studies evaluated the effects of granulocyte colony stimulating factor (G-CSF). The first demonstrates that hemopoietic failure after burn wound infection is not related to defective endogenous G-CSF release. Of concern, G-CSF administered in animal models impairs immunoglobulin synthesis after a burn injury and raises caution regarding its mechanism before proceeding to clinical trials.

One recent study found that a single allele in the tumor necrosis factor-alpha promoter region is associated with increased risk for sepsis. Other studies have documented an association between intranuclear NF-κB and systemic inflammatory response syndrome or multiple-organ failure. It is increasingly clear that both genetic effects and specific intracellular activation states drive the inflammatory response, and our ability to describe these in time and understand the mechanisms for manipulation will allow more precise interaction with this complex clinical problem.

References

1. Sessler CN, Shepherd W. New concepts in sepsis. Curr Opin Crit Care (United States) Oct 2002; 8(5):465-72.
2. Ely EW, Laterre PF, Angus DC et al. Drotrecogin alfa (activated) administration across clinically important subgroups of patients with severe sepsis. Crit Care Med (United States) Jan 2003; 31(1):12-9.

Pneumonia in the ICU

Background

Pneumonia remains a significant source of morbidity and mortality in surgery. Identification and treatment of hospital acquired pneumonia has evolved over the past decade but optimal management continues to change as rapidly as the ICU flora.

Epidemiology

More than 300,000 cases of nosocomial pneumonia are reported each year with an associated mortality rate of 20-50%.[1] Risk factors for the development of pneumonia include subclinical aspiration of bacteria, trauma, surgical stress, underlying pulmonary dysfunction and exposure to hospital gram-negative organisms. Patients requiring tracheal intubation and ventilation have incidences of nosocomial pneumonia as high as 68%.

Naziri et al described clinical and immunologic alteration in patients with pneumonia. Their results found a mean time to onset from ICU admission was 5 days, the average temperature was 101.4°F, and a leukocytosis of 16,000 cells/mm^3.[2] The PaO$_2$ was maintained > 100 at mean FiO$_2$ of 0.47. Patients with poor outcomes had a greater temperature elevation and leukocytosis after 5 days of pneumonia.

Diagnosis

The diagnosis of pneumonia at the University of Louisville has been by identification of 3 of 5 clinical markers. These are new onset of fever, decreasing oxygenation, purulent sputum, chest radiographic changes, and/or leukocytosis.

Accurate identification of the causative organism is frequently elusive, however, several methods have found their way into routine use. Bronchioalveolar lavage (BAL) and protected specimen brushings (PSB) provide the most accurate identification of the causative organism. Sputum gram stain is used to guide empiric therapy despite their reported unreliability in ventilated patients.

Bronchioalveolar lavage and protected specimen brushings have a wide sensitivity and specificity due to a lack of a consensus gold standard for comparison. Quantitative as well as qualitative cultures have been effective in assessing organisms. The current recommendation for a positive BAL is 10^5 cfu/mL or PSB of 10^3 cfu/mL.

Microbiology

Gram negative organisms are responsible for 75-90% of nosocomial pneumonias. The source of these are from the oropharynx and intestinal tract. The upper respiratory tract becomes colonized with gram negative bacteria in the first 48 hours of intubation. This contributes to confusion between colonization of the airway versus true infection.

Management

Aggressive pulmonary care in the ICU cannot be over emphasized. Frequent suctioning and therapeutic bronchoscopy are of significant clinical utility. Suspected nosocomial pneumonia should be treated with empiric antibiotics after attempts are made at identification of the causative organism. The choice of antibiotics should be guided by the flora typically isolated in the ICU. Coverage for gram negatives including Pseudomonas is usually selected. Antibiotics such as broad spectrum penicillins (piperacillin, ticarcillin) are instituted with conversion to a narrow spectrum antibiotic once sensitivities are available.

Several studies have found monotherapy comparable to combination therapy. Newer broad spectrum single agents provide equivalent coverage while sensitivities are ending. Studies have also suggested that in the trauma patient multidrug therapy is associated with increased drug resistance.

References

1. Naziri W, Cheadle WG, Pietsch JD et al. Pneumonia in the surgical intensive care unit. Immunologic keys to the silent epidemic. Ann Surg 1994; 219(6):632-640.
2. Polk HC, Livingston DH, Fry DE et al. Treatment of pneumonia in mechanically ventilated trauma patients. Results of a prospective trial. Arch Surg 1997; 132(10):1082-1092.
3. Civetta JM, Taylor RW, Kirby RR. Critical Care. 3rd ed. PA, Philadelphia: Lippencott-Raven Publishers, 1997.

Adult Respiratory Distress Syndrome (ARDS)

Background

ARDS was first described by Rene Laennec in 1821 as "idiopathic anasarca of the lungs." Injured soldiers succumbing to respiratory failure were described as having a posttraumatic massive pulmonary collapse around World War I, while "shock lung" and "white lung" and "DaNang lung" all came during later military conflicts. It was not until the seminal description of ARDS by Ashbaugh and colleagues in 1967 that physicians recognized their differing descriptions of the same syndrome.[1]

Acute lung injury (ALI) and acute respiratory distress syndrome (ARDS) represent a single pathologic process of acute respiratory failure. Acute lung injury is characterized by:

Table 7.1. Disorders associated with ARDS

Disease	Incidence of ARDS
Sepsis	41%
Aspiration pneumonia	22%
Pulmonary contusion	22%
Massive transfusions	36%

Pepe PE, Potkin RT, Reus DH et al. Clinical predictors of the adult respiratory distress syndrome. Am J Surg 1982; 144:124-30.

- bilateral infiltrates on chest radiographs
- pulmonary capillary wedge pressure of 18 mm Hg or less
- absence of clinically evident left atrial hypertension
- ratio of partial pressure of arterial oxygen to the fraction of inspired oxygen (PaO_2:FiO_2) of 300 or less

The definition of ARDS is the same with a PaO_2:FiO_2 ratio of 200 or less with mortality rates of 34-60%. Most deaths are attributable to pneumonia, sepsis, or multiorgan dysfunction.

The combined incidence of ALI and ARDS is estimated to be 75 cases/100,000/year. Of patients admitted to the ICU, 2-3% will develop ALI or ARDS.

Etiology

Several clinical disorders are associated with the development of ARDS, including pneumonia, aspiration, pulmonary emboli, near drowning, inhalation injury, reperfusion pulmonary edema, trauma, surgery, burn injury, drug overdose, acute pancreatitis, and massive blood transfusions. Overall, sepsis and aspiration pneumonia are associated with the highest risk of progression to ARDS, with 18-40% of sepsis cases developing ARDS (Table 7.1).

Pathophysiology

It is driven by the initiation of the systemic inflammatory response syndrome (SIRS). This may be infective as in sepsis or noninfective as in pancreatitis or a result of multisystem organ failure. ARDS progresses through an inflammatory, exudative, proliferative and fibrotic phase. Neutrophil activation as well as the production of TNF, IL-1, IL-6 and IL-8 lead to cell injury and death. This results in an inflammatory infiltration of the lung with altered capillary permeability and accumulation of proteinaceous pulmonary edema manifest as increased extravascular lung water. Frequently hypoxic vasoconstriction is lost in the ventilated patient when 100% FiO_2 is utilized. Ultimately pulmonary hypertension is the end result due to vasoconstrictors as well as microvascular thrombosis.

Diagnosis

As notes in the introduction the diagnosis of ARDS lies along a spectrum with the parameters heralding its onset.

- bilateral infiltrates on chest radiographs
- pulmonary capillary wedge pressure of 18 mm Hg or less

- absence of clinically evident left atrial hypertension
- ratio of partial pressure of arterial oxygen to the fraction of inspired oxygen ($PaO_2 : FiO_2$) of 200 or less

Management

Prone Positioning in ARDS

It has been shown that prone positioning improves oxygenation for 60% to 70% of ALI/ARDS patients, though the patients most likely to respond are not readily identified in advance. The mechanisms responsible for oxygenation improvements are uncertain, however, primary mechanisms likely responsible are:

- improved ventilation-perfusion matching
- changes in lung mass and shape
- alterations in compliance

Ventilation-perfusion matching may be severely impaired in patients with ALI/ARDS, partially related to their inability to normally produce hypoxic pulmonary vasoconstriction. In turning a patient from supine to prone, more homogeneous gas distribution is achieved, thus contributing to improved ventilation-perfusion matching. Also, in the process of turning from supine to prone, the heart is no longer compressing the posterior aspects of the left lung, allowing for better aeration. Furthermore, the majority of lung tissue is posterior, and prone positioning allows this large amount of lung tissue to function as anterior regions with better aeration (though perfusion may not improve or worsen).

Concerns regarding prone positioning remain, primarily related to the safety of performing the prone maneuvers and providing patient care for patients in this position. Of 152 patients enrolled prone positioning oxygenation improved in approximately 70% of prone positioning patients, and no significant difference was noted in adverse events related to the prone positioning.

High concentrations of inspired oxygen should be minimized in the patient with ARDS. In most cases following the SaO_2 to maintain a level > 90% is satisfactory. Levels less than 88% result in a steep decline in delivery as predicted by the oxygen-hemoglobin dissociation curve.

Lung Recruitment Maneuvers in ARDS

The predominant physiologic effect recognized from recruitment maneuvers is the increase in functional residual capacity of the lung, representing additional alveoli being opened and available for gas exchange. The most common method of recruitment includes PPEP and inverse ratio ventilation.

The Role of High-Frequency Ventilation (HFV) in ARDS

HFV (RR>60) may be defined as mechanical ventilation applied with high respiratory rates and low tidal volumes. There are a number of forms of HFV, including jet (100-200 Hz) ventilation and oscillatory ventilation. Early studies of HFV demonstrated improvements in oxygenation and in pathologic changes of lung injury or ARDS. Trials that have attempted to study the clinical effectiveness of HFV in ALI/ARDS patients have failed to observe any improvements. This may relate to pre-existing and ongoing lung injury, related to conventional ventilation strategies before the patient began HFV.

Use of Low Tidal Volume Ventilation

Traditional tidal volumes of 10-14 mL/kg can result in significant ventilator associated lung injury. Several studies have documented that low tidal volume ventilation, or permissive hypercapnea, is the only strategy proven to improve survival in ALI/ARDS patients. Mortality was decreased from 40% to 31%. Respiratory acidosis is tolerated well up to a pH of 7.20. Most institutions would reserve bicarbonate infusion for levels pH<7.10. There are many physiologic effects to be considered with respiratory acidosis and hypercarbia. These include myocardial depression, systemic vasodilatation, rightward shift of the oxygen-hemoglobin dissociation curve. This results in increased cardiac output, heart rate and decreased systemic vascular resistance. Relative contraindications to permissive hypercapnea include closed head injuries with increased intracranial pressures.

Strategies of ventilator support include lower tidal volumes in the range of 4-8 mL/kg as well as pressure limited settings.

Liquid Ventilation

ARDS is associated with a decrease in surfactant and increased alveolar surface tension. It has been theorized that the use of liquids with oxygen carrying capacity could be used to open and oxygenate alveolar spaces. The use of perfluorocarbons has been initiated in some trials. Total liquid ventilation has proven difficult and expensive opening the door to the use of partial liquid ventilation. In this process the lung is filled to FRC and ventilated with a conventional ventilator. Studies have shown it to be safe but no RCTs have shown it to be more effective.

References

1. Ashbaugh DG, Bigelow DB, Petty TL et al. Acute respiratory distress in adults. Lancet 1967; 2:319-323.
2. Cordingley JJ, Keogh BF. The pulmonary physician in critical care 8: Ventilatory management of ALI/ARDS. Thorax 2002; 57:729-734.

CHAPTER 8

Surgical Oncology
William R. Wrightson

Surgical Oncology Principals

Background
Cancer is responsible for over 500,000 deaths annually. It accounts for one quarter of all deaths in the US each year (Table 8.1).

Epidemiology
Basic principles regarding population studies of cancer include the incidence of a cancer and its prevalence. The incidence is the number of new cases reported in a population in a given year while the prevalence relates to the total number of diagnosed cases each year. Cancer is multifactoral in most instances and includes the interaction of genetic and environmental influences. Carcinogens may be in the form of physical agents (radiation), chemical agents (tobacco), and viral agents (hepatitis B virus). Ultimately, alterations in gene expression are the basis for all cancer.

Tumor Biology
Cells maintain homeostasis through a regulation of cellular proliferation and degradation. In addition a strict regulation of their genetic sequences to maintain their integrity is essential to inhibit mutagenesis and transformation.

Cancer is the result of a deregulation of these cellular functions where proliferation is greater than cell death. Cancer is unregulated cell growth.
- angiogenesis
- cell adhesion
- migration
- metastasis

Pathology
Neoplasia literally means new growth and is commonly used to describe malignancies. Hyperplasia reflects an increase in cell number. This can be seen in both normal and neoplastic tissues. Transformation of hyperplastic tissue into frank neoplasia is rare. Metaplasia represents a reversible transformation of one cell type to another. This is seen with Barrett's mucosa with a change from normal squamous epithelium to an intestinal columnar type. This is frequently the result of chronic irritation and inflammation of the epithelium. Dysplasia refers to an alteration in size, shape and organization of adult cells.

The dominant feature of a carcinoma is its ability to invade the basement membrane and metastasize to surrounding and distant tissues. Local disease is restricted to the primary site, regional disease to the surrounding tissues including lymph nodes and distant disease is metastatic to distant organ sites.

Current Concepts in General Surgery: A Resident Review, edited by William R. Wrightson.
©2006 Landes Bioscience.

Table 8.1. Relative frequency of cancer by sex

Cancer	Men	Women
Lung	32	25
Prostate	14	-
Breast	< 1	17
Colon	9	10
Leukemia/lymphoma	9	8
Pancreas	5	5

Tumor cells typically require 30 doublings to acquire a size of 1 cm and thereby become palpable.

Genetic Alterations

Cancer results from the loss of normal growth regulation. There are multiple check points in the cell cycle that control progression and thus proliferation. There are also pathways that control cell death or apoptosis. Two classes include oncogenes and tumor suppressor genes.

Treatment Principles

The primary goal of treatment is to eliminate tumor cells locally and regionally. There are three available means of accomplishing this goal. They are surgery, chemotherapy or radiation therapy.

8

Molecular Staging

Background

The most significant prognostic factor in most solid tumors is regional lymph node status. Patients with nodal metastasis in melanoma have a 40% decrease in 5 year survival as compared to node negative patients. Evaluation of nodal basins in melanoma has evolved in recent years with the advent of sentinel node biopsy techniques. First described by Morton et al, the sentinel lymph node (SLN) is the first node in a regional basin that receives cutaneous lymphatic afferents from a primary tumor. It is believed the sentinel node, whether positive or negative for tumor, can accurately predict the status of the remaining nodes.

One important factor in determining the status of the nodes is the method of pathologic evaluation. Standard pathologic evaluation takes one to two sections from the center of the node with routine H&E staining and histologic evaluation. This studies less than 1% of the total nodal tissue. Other studies have shown that routine histologic evaluation of regional nodes can fail to detect as high as 50% of metastatic disease. This suggests the presence of micrometastatic disease that was missed by routine histopathology. With SLNB, one or two nodes most likely to contain metastasis are submitted for evaluation. This allows for a more detailed examination of the tissue with serial sectioning, immunohistochemical staining and molecular staging techniques. However, specimens from a completion lymph node dissection are typically examined with standard histopathology and may miss micrometastatic disease.

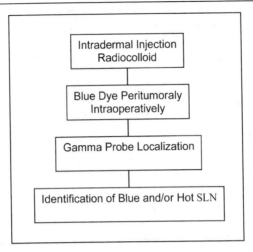

Figure 8.1. Sentinel node biopsy technique.

Molecular Staging

Molecular staging utilizes markers specific to a tumor of interest to minimize missed submicroscopic disease and therefore false negatives. Tyrosinase is an enzyme involved in the synthesis of melanin in normal melanocytes. It is specific to melanoma cells and is not expressed in normal lymph nodes. Use of reverse transcriptase polymerase chain reaction (RT-PCR) to detect tyrosinase mRNA and has been applied in many studies as a marker for melanocytes or melanoma cells. Patients that are histologically positive for melanoma are RT-PCR positive in 94-100% of cases. RT-PCR has also been shown to upstage 13-65% of patients with histologically negative nodes were RT-PCR positive.

The vast majority of melanoma patients with histologically positive sentinel lymph nodes (SLN) have no additional positive lymph nodes detected upon completion lymph node dissection (CLND). This information suggests that CLND may not be required for all patients with positive SLN. This study was conducted to determine the presence of micrometastasis with RT-PCR in the CLND specimen in those patients with histologically positive SLN and histologically negative nonSLN.

Sentinel Node Biopsy

Background

There continues to be significant debate with regard to the optimal technique for sentinel node biopsy (SNB), and it is the topic of study in many clinical trials. Since its introduction for melanoma and subsequently breast cancer, SNB has been applied with an array of techniques to facilitate its accurate identification. Initially, SNB was described with blue dye identification.[1] Subsequent studies investigated the use of radiocolloid and finally a combination of the two was utilized.[2,3] In addition, optimal application of these tracers has influenced SNB. Sifting through the techniques reported in current literature, we have identified a method that appears to consistently identify the SLN while minimizing the false negative rate (Fig. 8.1).

Table 8.2. Published studies of breast lymphatic mapping and sentinel lymph node biopsy

Author	Date	Number of Patients	SLN Identified N (%)	Technique	Pos SLN	Pos Axillary LN	False Neg Rate
Giuliano et al[1]	1994	174	114 (66%)	Blue dye alone	37	42	11.9%
Krag et al	1996	70	50 (71%)	Radiocolloid alone	21	21	0%
Albertini et al[3]	1996	62	57 (92%)	Blue dye plus radiocolloid	18	18	0%
Giuliano et al[10]	1997	107	100 (93%)	Blue dye alone	42	42	0%
Veronesi et al[5]	1997	163	160 (98%)	Radiocolloid alone (subdermal injection)	81	85	4.7%
Barnwell et al[18]	1998	42	38 (90%)	Blue dye plus radiocolloid	15	15	0
Krag et al[19]	1998	443	413 (91%)	Radiocolloid alone			11%
Flett et al[20]	1998	68	56 (82%)	Blue dye alone	53	56	5%
Bass et al[21]	1999	700	665 (95%)	Blue dye plus radiocolloid	238		0.84%

8

Technique

Use of blue dye, radiocolloid or a combination of the two appears to yield similar results (Table 8.2); however we believe both agents optimize locating the SLN. Although some surgeons have had good results without using blue dye at all, there is no disadvantage to using the blue dye, and we feel that it is a helpful technique that is complementary to the radiocolloid injection.

Intradermal Radiocolloid Injection

Current recommended guidelines for performance of radioguided sentinel lymph node biopsy include intradermal injection of 0.5 mCi of 0.2 micron filtered technetium-99 sulfur colloid in a volume of 6 mL at least 1 h prior to operation. For intradermal injection, filtered technetium sulfur colloid is injected in four or five locations into the skin (raise a wheal) overlying the tumor. For tumor locations in which it is not possible to identify the area of skin overlying the tumor, peritumoral injection is recommended. It is important to use the larger volume (6 mL) for peritumoral injection. However, the smaller volume for skin injection is preferable. Intradermal injection appears to reflect accurately the lymphatic drainage of the breast tissue beneath it and results in much more radioactivity reaching the sentinel nodes. Therefore, sentinel node identification is often easier. The other advantage is that the radiocolloid is not diffused throughout the breast tissue, which in upper outer quadrant lesions can often lead to a very difficult time in identifying the sentinel lymph nodes, especially percutaneously.

It has been shown that preoperative lymphoscintigraphy is not necessary in the preoperative evaluation and adds significant expense. In a series of patients we found no difference in the identification of the SLN with or without lymphoscintigraphy. We no longer recommend its routine use in breast SLN localization.

Intraoperative Injection with 5 mL Iymphazurin Dye

Blue dye is injected peritumoral with of 5 mL lymphazurin dye (1% isosulfan blue). The dye should be injected within 5-10 minutes of making the incision to search for the sentinel nodes. Gentle massage of the breast after injection for 5 minutes seems to speed the flow of dye through the afferent lymphatics.

There are several options for injection of the blue dye. For palpable tumors, the dye may be injected around the palpable tumor directly. Ultrasound guidance may be used to inject the dye around the tumor or biopsy cavity if prior excisional biopsy was performed. For nonpalpable lesions for which wire localization has been performed, the entire 5 mL of dye may be injected down the needle, which has been left in place.

Alternatively, the blue dye may be injected under direct vision. If lumpectomy is performed, the dye can be injected into the normal breast tissue following removal of the specimen. For women who have had an excisional biopsy and have elected to have a mastectomy, the previous biopsy cavity can be opened and the vital blue dye injected under direct vision. The biopsy cavity should then be closed and gloves and instruments changed.

The location of the sentinel node should be determined, if possible, by transcutaneous scanning using a hand-held gamma counter. A "hot spot" in the axilla indicates the location of the sentinel node.

Sentinel Node Biopsy

A small incision will be made in the axilla over the suspected location of the sentinel node. If no hot spot is identified, a curved transverse (anterior to posterior) incision in the lower axilla will give excellent exposure. After incising the clavipectoral fascia to gain access to the axillary contents, the gamma counter is again used to pinpoint the location of the sentinel node. If the tumor is in the upper outer quadrant, the background radioactivity from the tumor injection may make it difficult to detect the sentinel node. Blue stained afferent lymphatic channels should be sought and dissected to the blue-staining sentinel nodes.

All blue lymph nodes should be removed. Similarly, any node with significant radioactivity over background should be removed. Usually, the blue nodes are also the most radioactive, indicating that both techniques identify the same sentinel node. It is possible; however, to have radioactive lymph nodes that are not blue, or blue lymph nodes that are not excessively radioactive. The number of counts per second (cps) should be recorded for each sentinel node in situ, and after removal from the patient. The latter is best performed by placing the node on top of the gamma probe, pointed toward the ceiling.

After removal of each sentinel node, the background activity in the axilla should be examined with the probe and recorded. Background activity less than 10 times the activity of the hottest lymph node indicates that all sentinel nodes have been removed. If no sentinel node is identified, axillary dissection should be carried out in standard fashion.

Each sentinel node should be placed in formalin and labeled separately as sentinel node #1, #2, #3, etc. It is also helpful to record the number of cps for each sentinel node on the pathology sheet.

After the sentinel node(s) are removed, completion level I and II axillary lymph node dissection should be performed. This includes the lymph nodes lying beneath the pectoralis minor muscle, to the axillary vein superiorly, the latissimus dorsi laterally, and the serratus anterior medially. The long thoracic and thoracodorsal nerves should be identified and preserved. The axillary dissection specimen should be sent separate from the sentinel nodes and clearly marked.

Pathology

Because we know that more intensive pathologic investigation of the sentinel nodes will detect micrometastatic disease and that such micrometastases correlate with worse prognosis, it makes little sense to perform sentinel lymph node biopsy and submit the nodes for routine histology. Each sentinel node should be processed by cutting the lymph node at 2-3 mm intervals ("bread loafing") for paraffin embedding of each piece. At least one section should be taken from each piece (about 5 sections for a 1.0 cm lymph node). For smaller lymph nodes, multiple pieces can be embedded in a single block. Routine H&E stains are performed. The axillary dissection specimen should be examined using routine histologic methods.

References

1. Giuliano AE, Kirgan DM, Guenther JM et al. Lymphatic mapping and sentinel lymphadenectomy for breast cancer. Ann Surg 1994; 220:391-401.
2. Krag DN, Weaver DL, Alex JC et al. Surgical resection and radiolocalization of the sentinel lymph node in breast cancer using a gamma probe. Surg Oncol 1993; 2:335-340.
3. Albertini JJ, Lyman GH, Cox C et al. Lymphatic mapping and sentinel node biopsy in the patient with breast cancer [see comments]. JAMA 1996; 276:1818-22.
4. Veronesi U, Paganelli G, Galimberti V et al. Sentinel-node biopsy to avoid axillary dissection in breast cancer with clinically negative lymph-nodes. Lancet 1997; 349:1864-67.
5. Morton DL, Wen D-R, Wong JH et al. Technical details of intraoperative lymphatic mapping for early-stage melanoma. Arch Surg 1992; 127:392-399.
6. Alex JC, Krag DN. The gamma-probe-guided resection of radiolabeled primary lymph nodes. Surg Oncol Clin N Am 1996; 5:33-41.
7. Giuliano AE, Kirgan DM, Guenther JM et al. Lymphatic mapping and sentinel lymphadenectomy for breast cancer. Ann Surg 1994; 220:391-8; discussion 398-401.
8. Giuliano AE, Dale PS, Turner RR et al. Improved axillary staging of breast cancer with sentinel lymphadenectomy. Ann Surg 1995; 222:394-9; discussion 399-401.
9. Giuliano AE, Jones RC, Brennan M et al. Sentinel lymphadenectomy in breast cancer. J Clin Oncol 1997; 15:2345-2350.
10. Giuliano AE. Sentinel lymphadenectomy in primary breast carcinoma: An alternative to routine axillary dissection [editorial]. J Surg Oncol 1996; 62:75-7.
11. Nathanson SD, Anaya P, Karvelis KC et al. Sentinel lymph node uptake of two different technetium-labeled radiocolloids. Ann Surg Oncol 1997; 4:104-10.
12. Statman R, Giuliano AE. The role of the sentinel lymph node in the management of patients with breast cancer. Adv Surg 1996; 30:209-21.
13. Rosen PP, Lesser ML, Kinne DW et al. Discontinuous or "skip" metastases in breast carcinoma. Analysis of 1228 axillary dissections. Ann Surg 1983; 197:276-83.
14. Van Lancker M, Goor C, Sacre R et al. Patterns of axillary lymph node metastasis in breast cancer. Am J Clin Oncol 1995; 18:267-272.
15. Dowlatshahi K, Fan M, Snider HC et al. Lymph node micrometastases from breast carcinoma. Reviewing the dilemma. Cancer 1997; 80:1188-1197.
16. Clare SE, Sener SF, Wilkens W et al. Prognostic significance of occult lymph node metastases in node-negative breast cancer. Ann Surg Onc 1997; 4:447-451.
17. Burak, William Jr E, Walker et al. Routine preoperative lymphoscintigraphy is not necessary prior to sentinel node biopsy for breast cancer. Am J Surg 1999; 177(6):445-449.

8

CHAPTER 9

Breast

Sara Elizabeth Snell

Breast Cancer Screening and Diagnosis

Screening

Screening tools commonly used include the patient's history, self and physician exam, mammography and ultrasound.

Mammogram recommendations:

1. Baseline mammograms are recommended in the United States at age 35.
2. Every year beginning at age 40.
3. In high risk patients, mammograms are recommended 5 years prior to the first case of breast cancer in the close family.[1]

Limitations in mammography are secondary to dense breast tissue overlying lesions frequently seen in women <30 years old and older women on hormone replacement.[2] Mammogram is associated with a false negative rate of 10-15%. This false negative rate is increased to 25% in women less than 40 years old.

The benefit of mammogram is the detection of lesions that are not yet palpable. Screening by mammogram can detect groups of microcalcifications that indicate ductal carcinoma in situ (DCIS). If the microcalcifications are linear or ductal, this indicates comedo type which has a worse prognosis. If they are sand-like, then cribiform or papillary is more common. These latter two have a better prognosis. This allows mammogram to detect cancer prior to invasion.

Ultrasound is also used to determine if a palpable lesion is cystic or solid. Benign lesions are very circumscribed. Malignant lesions are asymmetric with a lesion that is taller than it is wide and with irregular borders. Malignant lesions are hypoechoic and have posterior shadows. This technology is very helpful in women with dense breasts. The limitation with ultrasound is that microcalcifications are not viewed. Another limitation is that the technique is operator dependent and time consuming. Some cancers are visualized on ultrasound that are not seen on mammogram.

MRI is not used as a screening tool, but is used to determine recurrence vs. scar in lumpectomy patients. Recurrences have increased vascularity as seen on MRI.

Diagnosis

Stereotactic Biopsy

New biopsy techniques involve what is termed as minimally invasive management. This either uses stereotactic or ultrasound guided biopsy.[3] Stereotactic machines biopsy microcalcifications or a nonpalpable mass on a geometric X, Y, and Z axis.[4] Core biopsies are obtained by at least a 14 gauge needle. At least five cores are needed to perform an adequate tissue sample.[5] A clip is usually placed at the biopsy site for future imaging and possible therapy.

Current Concepts in General Surgery: A Resident Review, edited by William R. Wrightson. ©2006 Landes Bioscience.

Limitations with this technique include a lesion close to the skin or nipple or a patient with small breasts or a patient that is uncooperative. A very important point with stereotactic biopsy is that a diagnosis of atypia or DCIS cannot be made.[6] This lesion needs to be openly excised to check for invasion.

An older option of diagnosis of nonpalpable lesion is needle localization by mammogram or ultrasound. The tissue removed should be sent to radiology or pathology to confirm removal of the lesion. This, however, requires an operative procedure and is more disfiguring to the breast.

Metastasis

Metastatic spread is by different pathways. The main metastatic area is to the ipsilateral axillary nodes as it drains 75% of the breast. Parasternal nodes drain the medial 25% of the breast. Vascular spread sends metastasis to the lungs, brain, and spine.

Preoperative evaluation includes:

1. LFT to check liver involvement
2. CXR to check for lung mets
3. Abdominal CT is performed if the LFTs are elevated
4. Bone scan is performed if there is elevated calcium or the patient has bone pain

If patients have metastatic disease to the bone, they may present with hypercalcemia. The treatment of this involves hydration, a loop diuretic like Lasix, and a long term biphosphonate (pamidronate) IV. If the patient presents with metastatic disease, no surgery is preformed and comfort measures only are invoked. If spinal cord compression syndromes present then dexamethasone is given and the patient may require spinal stabilization.

References

1. Dershaw DD. Mammographic screening of the high-risk woman. AJS 2000; 180:288-289.
2. Lucassen A, Watson E, Eccles D. Advice about mammography for a young woman with a family history if breast cancer. BMJ 2001; 1040-1042.
3. Smith DN, Christian R, Meyer JE. Large-core needle biopsy of nonpalpable breast cancers. Arch Surg 1997; 132:256-259.
4. Dershaw DD, Liberman L. Stereotactic breast biopsy: Indications and results. Oncology 1998; 12(6):907-922.
5. Rich PM, Michell MJ, Humphreys S et al. Stereotactic 14 G core biopsy of nonpalpable breast cancer: What is the relationship between the number of core samples taken and the sensitivity for detection of malignancy? Clin Rad 1999; 54:384-389.
6. Brem DF, Behrndt S, Sanow et al. Atypical ductal hyperplasia: Histologic underestimation of carcinoma in tissue harvested from impalpable breast lesions using 11-gauge stereotactically guided directional vacuum-assisted biopsy. AJR 1999; 172:1405-1406.

9

Breast Cancer: Benign Breast Disease

Anatomy

The breast is composed of glandular tissue called lobules. Each of these lobules is drained by 15-20 lactiferous ducts in each breast.

The blood supply of the breast is provided by lateral perforating branches of the internal thoracic artery, also known as the internal mammary artery. In 75% of patients, the lateral blood supply of the breast is based on perforating medial branches of the lateral thoracic artery. Medial breast lymph drainage is to the parasternal lymph nodes. Lateral breast lymph drainage is by the axillary nodes.

Multiple breasts can develop in both males and females, following a predetermined milk line. Multiple breasts are termed polymastia. If only an extra nipple or areola is present, this is termed polythelia. The unilateral absence of a breast is referred to as Poland's syndrome. Poland's syndrome can also be associated with the absence of the pectoralis muscle.

Benign Breast Disease

The initial patient presentation is generally with a complaint of breast pain (mastalgia) or breast lump.

Mastalgia

Rarely does breast pain represent cancer; however a thorough breast exam and evaluation should be performed. Breast pain is characterized as cyclical or noncyclical with treatment as determined by the type of pain present. Cyclic pain is probably hormonal in nature, since it can be associated with elevated prolactin levels and relieved with menopause. Noncyclic mastalgia affects older women and the origin of the pain should be discriminated as chest wall or breast pain.

Management

Treatment can often be with a support bra worn at night. If conservative treatment fails, medicine would be instituted. Medication must be continued for at least two months before being considered a treatment failure. The drug with the least side effects and an average success rate at decreasing pain is gamolenic acid or evening primrose oil. Other medications with efficacy include danazole, a synthetic testosterone (inhibits the gonadotropin surge and enzymes of steroid synthesis) and bromocriptine (a dopamine antagonist).

Breast Lump

Identification of a breast lump is the most common reason patients seek medical attention. Breast lumps in women less than 40 are often benign but should be assessed in a methodical fashion.

Diagnosis

Often now referred to as the triple test, physical exam, mammogram and biopsy are the initial diagnostic methods. Sometimes ultrasound can be used to evaluate a questionable mass. MRI and PET scans are being considered as tools; however, now they are cost prohibitive and have questionable sensitivity and specificity. If the mass is still questionable, then core needle or excisional biopsy can be preformed.

9

Fibroadenoma

Masses in women less than 40 are most often a fibroadenoma, which develops from abnormal lobule development. Fibroadenomas are composed of stromal and epithelial elements. On a physical exam, these are discrete masses that are freely movable and soft.

Diagnosis

On an ultrasound, they are solid masses that have well circumscribed margins with a heterogeneous echo pattern. On a mammogram, a mass with circumscribed borders is seen. Four types of fibroadenomas exist:

1. Common
2. Giant
3. Juvenile
4. Phyllodes tumor

Giant fibroadenomas are ones that measure greater than 5 cm. Juvenile fibroadenomas are also large and occur in adolescent girls. Phyllodes tumors are seen in premenopausal women. Treatment of phyllodes tumors are with excision; however, they tend to recur. Fibroadenoma will increase in size in 10% of cases while 33% will regress.

Fibroadenomas are a long term risk factor for cancer only if proliferative disease is present or the fibroadenoma is complex.[3] A complex fibroadenoma is one that contains cysts, sclerosing adenosis, epithelial calcifications, or apocrine changes.

A biopsy should be done if:

1. increases in size on ultrasound
2. ≥ 3 cm or greater in size
3. atypia on x-ray or previous biopsy

Fibrocystic Disease

Fibrocystic disease is more common in women around age 40. This disease was first described in 1829 by Astley Cooper, of Cooper's ligament fame. These cysts form secondary to involution of the breast lobules.

Diagnosis

One to three per cent of cysts have been associated with breast cancer displayed by internal shadows as seen on ultrasound. Abnormal borders are sometimes seen on mammogram. These abnormal cysts should be aspirated and, if bloody, sent for cytology. Cysts should be excised for the following reasons:

1. aspirate is bloody
2. residual mass exists after aspiration
3. persistent refilling after aspiration

Cysts that carry an increased risk of cancer are ones that are palpable or ones that develop in women at a young age.[4] Cysts that have atypical hyperplasia also have an increased risk of cancer.[3]

Other Benign Masses

There are other less common benign masses in women. These include lipomas, found in women of all ages. They are soft, fleshy and mobile and are usually diagnosed by excision. Fat necrosis is another benign mass, resulting from trauma. Fat undergoes involution to become a hard palpable mass. Because this mass is hard, distinguishing it from a malignant tumor is difficult. Diagnosis is often made by excision. Lymph nodes in the breast are sometimes palpable, but a mammogram can usually determine this to be a benign lesion.

Breast Abscess

Breast infection sometimes is present as a palpable indurated mass. These are seen in nonlactating and lactating women but are far more common in the latter. Early treatment with antibiotics can prevent abscess formation. Treatment of breast abscess involves drainage of the collection. This can be done with incision and drainage or repeated aspiration and antibiotics, usually nafcillin or amoxicillin, to cover *S. aureus*.[5] More extensive and loculated collections may be aspirated with ultrasound guidance.[6]

If an underlying lesion is present after treatment, then biopsy must be performed to rule out carcinoma. If the lesion is solid at the first aspiration, biopsy must be performed to rule out inflammatory carcinoma.

Treatment of a lactating abscess also includes the continued drainage of milk of the affected segment. Antibiotics that cannot be used in breast feeding mothers include floxins, which causes abnormal cartilage formation; sulfas, an increase in free bilirubin by displacing it from albumin and resulting in secondary kernicterus; or tetracyclines creating abnormal teeth development.

Nonlactating abscesses can occur periareolar or peripherally. Periareolar abscess occur more often in young women that smoke.[1] This, pathologically, is seen as inflammation around nondilated subareolar ducts and can progress to a mammary duct fistula creating a communication between the skin and a subareolar duct. This fistula is usually seen after incision and drainage of an abscess. Treatment is by excision of the fistula and duct and administrating antibiotics. Peripheral nonlactating abscesses are associated with immune compromise. They are seen in diabetics or women with chronic steroid use. Incision and drainage is only done if the overlying skin is compromised.

Nipple Discharge

Another benign complaint is nipple discharge. If the discharge is bilateral, a non-breast source is often the cause. This may indicate an increased prolactin level. This suggests a pituitary tumor or a medication side effect (e.g., an anti-psychotic drug [dopamine blocker] can also stimulate this). Nipple discharge is only worrisome if it is bloody or spontaneously drains from one duct. This can be a sign of an intraductal papilloma or an invasive cancer. The most common cause of bloody nipple discharge is a benign intraductal papilloma. Investigation usually requires a ductogram to isolate the involved duct that, then, undergoes excision.

1. True or false: Breast abscesses are treated primarily with incision and drainage.

A: False. They can usually be treated with aspiration and IV antibiotics. If a residual mass is present is should be biopsied.

2. Periareolar abscesses are seen in which patients?

A: Smokers

References

1. Dixon JD, Morrow M. Breast disease a problem based approach, 1st ed. London: W.B. Saunders, 1999:1-205.
2. Mansel RE. Breast pain. BMJ 1994; 309:866-868.
3. Dupont WD, Page DL, Parl FF et al. Long-term risk of breast cancer in women with fibroadenoma. NEJM 1994; 331(1):10-15.
4. Dixon JM, McDonald C, Elton RA et al. Risk of breast cancer in women with palpable cysts: A prospective study. Lancet 1999; 353:1742-1745.
5. Dixon JM. Breast infection. BMJ 1994; 309:946-949.
6. Schwarz RJ, Shrestha RS. Needle aspiration of breast abscesses. Am J Surg 2001; 182:117-119.

Breast Cancer

Risk Factors and Genetics

Breast cancer risk factors are related to prolonged exposure to estrogen. This is seen in women with early menarche and late menopause, older high estrogen dose oral contraceptives, and nulliparity. The highest risk involves a personal history of breast cancer or lobular carcinoma in situ. Family history in a premenopausal first degree relative is also an important risk factor.

Familial breast cancer has been associated with certain genes. The two main genes are BRCA I and BRCA II.[1] BRCA I is found on chromosome 17 and is also associated with an increased risk of colon, ovarian, and prostate cancer. BRCA I is thought to function as a tumor suppressor gene in DNA repair. BRCA II, located on chromosome 13, is associated with male breast cancer. BRCA II is also thought to function as a tumor suppressor gene. Both of the above genes are inherited in an autosomal dominant pattern with incomplete penetrance. Prophylactic bilateral mastectomy has been shown to decrease breast cancer in women with the previous listed mutations in early studies.[2] Other syndromes have increased risk of breast cancer include Li-Fraumini syndrome and p53 mutations. The most common abnormally expressed gene in any breast cancer is p53 on chromosome 17. The risk of breast cancer is especially increased with exposure to radiation. Male breast cancer is increased in BRCA II and in Klinefelter's syndrome with gynecomastia.

Pathology

It is believed that breast cancer progresses from atypia, to ductal carcinoma in situ (DCIS), then to invasive cancer.

Lobular carcinoma in situ (LCIS) is associated with an increased risk of invasive cancer. This 15% of women with LCIS will develop an invasive carcinoma of ductal or lobular type in either breast at any site.[3] In contrast to DCIS, this is not a lesion that has a natural history to become an invasive cancer. Therefore excision of this lesion does not invoke protection against developing future cancer. The only treatment for this lesion involves bilateral mastectomy or close follow up. Bilateral mastectomy has been proven to decrease future carcinoma development.

Ductal carcinoma in situ (DCIS) is a precursor to invasive ductal carcinoma. This should be treated as a malignancy. An important distinction is made between DCIS and invasive cancer in that DCIS has an intact basement membrane. Treatment involves local control as either mastectomy or lumpectomy with or without radiation depending on lesion size and type.[4] A worse prognosis is associated with lesions > 2 cm, comedo type or excisional margin that is close to the lesion (< 1 mm). These are managed with postoperative radiation for local control. The disease can be multifocal in as many as 20% and in that case mastectomy is recommended. Axillary dissection is not routinely preformed for DCIS as it is not an invasive carcinoma and should not have metastatic spread. If the DCIS lesion is 5 cm or greater in size, invasion is assumed and the lesion is treated as so.

Infiltrating ductal carcinoma is the most common breast cancer. This is observed as a stellate lesion or architectural distortion on mammogram. In contrast, invasive lobular carcinoma is often not seen on mammogram and can lead to a delay in diagnosis. Any palpable lesion should be fully evaluated.

Paget's disease resembles a rash at the nipple that indicates underlying DCIS. If there is a palpable lesion at the nipple invasion is often present.

Inflammatory breast cancer presents as an inflamed, hard breast with diagnosis made after not responding to antibiotics. A biopsy will show invasion of dermal lymphatics. Plugging of these lymphatics cause the characteristic edema and peau d'orange seen in inflammatory cancer. This patient usually presents with metastatic disease and is labeled as a T3b lesion. Treatment of inflammatory cancer involves preoperative chemotherapy, for this has been shown to provide the best survival. If the patient survives chemotherapy, mastectomy follows and the radiation to the chest wall and sometimes axilla if gross disease is present.

Cystosarcoma phyllodes is the most common type of sarcoma of the breast. Sarcomas make up < 1% of all breast cancers. This is distinguished from a benign phyllodes tumor by the amount of cellular atypia present, mitotic activity (>10/HPF), and malignant infiltrating margins. No axillary dissection is performed because this is a sarcoma of mesoderm origin and spread by vascular routes not lymph.[5] The metastatic spread here is to the lungs and follow-up CT is usually preformed. These may recur at the primary site and require reexcision. There has been no proven advantage with chemotherapy or radiation pre- or postoperatively.

Staging
Tables 9.1, 9.2, 9.3 and 9.4.

Table 9.1. Breast cancer staging

Primary Tumor		
Tx		Primary tumor cannot be accessed
T0		No evidence of primary tumor
Tis		Carcinoma in situ
T1		Tumor less than 2 cm
	T1a	0.5 cm or less
	T1b	Greater than 0.5 cm but less than 1 cm
	T1c	Greater than 1 cm but less than 2 cm
T2		Tumor greater than 2 cm but less than 5 cm
T3		Tumor greater than 5 cm
T4		Tumor with direct extension into chest wall or skin
	T4a	Extension into chest wall
	T4b	Edema or ulceration of breast skin or satellite skin nodules
	T4c	Both T4a and T4b
	T4d	Inflammatory carcinoma

Table 9.2. Breast cancer regional node status

Regional Lymph Nodes

NX		Regional lymph nodes cannot be assessed
N0		No regional lymph nodes mets
N1		Mets to movable ipsilateral lymph nodes
	N1a	Only micromets (< 0.2 cm)
	N1b	Mets greater than 0.2 cm
N2		Mets to fixed ipsilateral lymph nodes
N3		Mets to ipsilateral internal mammary nodes

Table 9.3. Breast cancer metastasis

Distant Metastasis

MX	Presence cannot be assessed
M0	No distant mets
M1	Distant mets including ipsilateral supraclavicular nodes

Table 9.4. Breast cancer survival at 5 years

Stage	Description Survival %	5 year
Stage 0	Tis, N0, M0	> 95%
Stage I	T1, N0, M0	85%
Stage IIA	T0, N1, M0	70%
	T1, N1, M0	
	T2, N0, M0	
Stage IIB	T2, N1, M0	60%
	T3, N0, M0	
Stage IIIA	T0, N2, M0	40%
	T1, N2, M0	
	T2, N2, M0	
	T3, N1, M0	
	T3, N2, M0	
Stage IIIB	T4, Any N, M0	20%
	Any T, N3, M0	
Stage IV	Any T, Any N, M1	< 5%

9

References

1. Hill ADK, Doyle JM, McDermott EW et al. Hereditary breast cancer. Br J Surg 1997; 84:1334-1339.
2. Meijers-Heijboer H, van Geel B, van Putten WLJ et al. Breast cancer after prophylactic bilateral mastectomy with a BRCA I or BRCA II mutation. NEJM 2001; 345(3):159-164.
3. Frykberg ER. Lobular carcinoma in situ of the breast. Breast 1999; 5(5):296-302.
4. Skinner KA, Silverstein MJ. The management of ductal carcinoma in situ of the breast. Endo-Rel Ca 2001; 8:33-45.
5. Palmer ML, De Risi DC, Pelikan A et al. Treatment options and recurrence potential for cystosarcoma phyllodes. Surg Gyn Obs 1990; 170:193-196.

Management of Breast Cancer: Surgical Therapy

History
The radical mastectomy was first introduced by Halstead in the 1800s to treat local disease.

Surgical Options
The two main surgical options for invasive breast cancer in females are:
1. modified radical mastectomy
2. partial mastectomy

Partial mastectomy is also known as lumpectomy or quadrentectomy. These procedures are used for control of local and metastatic disease. Other surgeries, like simple mastectomy or subcutaneous mastectomy, are usually reserved for prophylaxis.

Modified radical mastectomy involves removal of breast tissue, nipple and skin overlying this along with the axillary contents. The axilla has three levels of nodes. Level I is lateral to the pectoralis minor muscle. Level II is beneath pectoralis minor and level III is medial to pectoralis minor. In mastectomy only levels I and II are taken as metastasis usually spreads from levels I to III in sequential order. There has been no increase in survival with taking level III, but it does increase lymphedema. Pectoralis muscle is only removed if the tumor directly invades. In radical mastectomy, a surgery that is no longer preformed, the pectoralis is routinely removed. However, this proved to be more disfiguring without survival advantage.

Partial mastectomy or breast conservation therapy (BCT), involves removal of the diseased tissue and obtains an uninvolved margin of greater than 1 cm. The goal is to:
1. Reduce the tumor burden to microscopic levels
2. Compliant patient for follow-up and radiation therapy for local control

In 1995 this was unequivocally proven equal to mastectomy in disease-free survival and overall survival, if followed with radiation for tumors less than 4 cm. Patients with DCIS were found to have an 8% ipsilateral recurrence rate with BCT and irradiation compared to 18% for BCT alone. Radiation after lumpectomy decreases local recurrence but does not improve overall survival. Breast conservation therapy is not indicated in T3 or T4 N2 MX lesions or lesions that are multifocal or multicentric. Certain other situations require mastectomy including collagen vascular disease, pregnancy, or a previously irradiated breast. Determination of nodal status should follow lumpectomy to determine the need for adjuvant chemotherapy. Positive axillary nodes increase the risk of metastatic recurrence but have no increased risk of local recurrence (Table 9.5).

Sentinel node biopsy can determine the status of the axillary nodal basin using minimally invasive surgical techniques. The advantages of sentinel node biopsy include decreased risk of nerve injury and post operative lymphedema. This postoperative lymphedema is a long term risk of angiosarcoma which presents as a blue dot on the involved arm. The disadvantage to omitting full dissection includes the advantage of local disease control in the axilla. For this reason, patients with grossly positive nodes are not candidates for sentinel node biopsy. Another issue with sentinel node involves determination of patients who need postoperative chemotherapy. Historically, patients with greater than four positive nodes received chemotherapy.

Table 9.5. Breast cancer studies

Study	N	Result
NSABP B-04	1765	In clinically node negative patients, total mastectomy with delayed node dissection, total mastectomy with radiation, and radical mastectomy are equivalent in disease free and overall survival
NSABP B-06	1843	Modified radical mastectomy, BCT with ALND and irradiation, and BCT with ALND were equivalent in terma odf overall survival. Breast irradiation did decrease breast tumor recurrents from 39% to 10% but did not affect overall survival.
NSABP B-17	818	Patients with DCIS receiving BCT and irradiation had a better 5 year survival compared to those receiving BCT alone. This was attributed to a higher incidence of invasive cancers in the BCT alone group (50% vs 29%).

Nerves injured in axillary dissection include the long thoracic to serratus anterior, thoracodorsal to latissimus dorsi and, more commonly, the intercostal brachial to the skin. Often times the latter nerve is sacrificed and a paresthesia to the upper arm develops.

References
1. Fisher B, Anderson S, Redmond CK et al. Reanalysis and results after 12 years of follow-up in a randomized clinical trial comparing total mastectomy with lumpectomy with or without irradiation in the treatment of breast cancer. NEJM 1995; 333(22):1456-1462.
2. Greenfield LJ, Mulholland M, Oldham KT et al. Surgery scientific principals and practice. 2nd ed. Philadelphia: Lippincott-Raven, 1997.

Breast Cancer: Adjuvant Therapy

Chemotherapy is traditionally initiated six weeks postoperatively. This allows the tissue time to heal prior to the immunosuppression associated with chemotherapy. Chemotherapy was historically reserved for patients with a poor prognosis. This included patients with greater than 4 positive nodes, primary tumors greater than 4 cm, a grade III tumor, or any tumor with lymphatic or vascular invasion. All of these imply systemic spread of tumor. Neoadjuvant chemotherapy is used in women who desire breast conservation but have large primary tumors. Preoperative chemotherapy has been shown to decrease the size of the primary tumor and, also, the incidence of positive nodes with no change in systemic recurrence.[1]

Chemotherapy can be divided into cytotoxic, hormonal, and molecular. Cytotoxic chemotherapy is the classic chemotherapy and includes four main drugs. The classic therapy is CMF (cyclophosphamide, methotrexate, and 5-fluorouracil). Adriamycin (doxorubicin) is sometimes added or substituted (Table 9.6).

Tamoxifen is the first hormonal chemotherapy drug and is used to block estrogen receptors.[2] This drug historically was used for tumors that are estrogen or progesterone receptor positive. Now, tamoxifen is used in women that are receptor positive or negative for prophylaxis against a new breast cancer. This is used after cytotoxic chemotherapy. Tamoxifen functions to put the cancer cells in G0 phase of cell division by blocking the estrogen receptor. Either used as prophylaxis against further

Table 9.6. Chemotherapeutic agents used in breast cancer

Agent Effect	Mechanism	Predominent Side
Cyclophosphamide	Alkylating agent that stops DNA replication	Hemorrhagic cystitis
Methotrexate	Acid analogue that inhibits DNA replication	Alopecia
5-Fluorouracil	Antimetabolite to inhibit DNA synthesis	Nausea and vomiting
Adriamycin	DNA antimetabolite that the exact function is unknown, but intercalates between DNA base pairs	Cardiomyopathy with a cumulative effect. Limited lifetime dose secondary to this side effect.
Taxol	Microtubule stabilizer preventing malignant cells from completing mitosis	Peripheral neuropathy

cancer or in receptor positive patients, no increased benefit has been seen with more than 5 years use. The side effect is an increased risk of uterine cancer secondary to the estrogen receptor blockade/stimulation.

Aromatase inhibitors are a new hormonal chemotherapy that serves to decrease end estrogen production. Aromatase is the enzyme that makes estrogen in a final biochemical pathway.

Molecular chemotherapy is being developed and one drug has been FDA approved. Herceptin is a monoclonal antibody to HER-2-neu. HER-2-neu is an epidermal growth factor receptor that has overexpression in poor prognosis breast cancer. This has been shown to decrease progression of disease in metastatic breast cancer.

References

1. Fisher B, Brown A, Mamounas E et al. Effect of preoperative chemotherapy on local-regional disease in women with operable breast cancer: Findings from national surgical adjuvant breast and bowel project b-18. J Clin Onc 1997; 15:7:2483-2493.
2. Early breast cancer trialists' collaborative group, tamoxifen for early breast cancer: An overview of the randomised trials. Lancet 1998; 351:1451-1467.

CHAPTER 10

Esophagus

William R. Wrightson

Barrett's Esophagus

History

Barrett's esophagus was first described by Norman Barrett in 1950 as an intestinal metaplasia of the distal esophagus.

Background

Adenocarcinomas constitute 2.5% to 8% of primary esophageal cancers, although this frequency is increasing dramatically in the United States at a rate surpassing that of any other cancer. It is suggested that these carcinomas develop from an increased incidence of Barrett's esophagus (BE). BE is metaplastic change of the normal squamous epithelium to an intestinal columnar type secondary to gastrointestinal reflux. The incidences of BE is 22/100000.

Patients with a columnar-lined lower esophagus (Barrett's metaplasia) are 40 times more likely to develop adenocarcinoma than the general population. Although the true incidence of Barrett's esophagus in the general population is unknown, it has been estimated that adenocarcinoma arises in 8% to 15% of patients with a columnar-lined esophagus. Patients with adenocarcinoma of the distal esophagus are discovered to have intestinal columnar metaplasia in 80% of cases. virtually synonymous with carcinoma in situ and being an indication for resectional therapy.

Etiology

BE results from severe gastroesophageal reflux diseases (GERD). It usually presents in the sixth decade of life with a male-to-female ratio of 3:1.

Anatomy

They occur most commonly in the distal third of the esophagus.

Pathophysiology

Refluxed gastric acid, proteases and bile erode normal squamous epithelium with residual pluripotential basal cells differentiate along varying lines. A wide variety of genetic events and mechanisms appear to play a role in the progression of Barrett's to adenocarcinoma. .Development of Barrett's is associated with chromosomal loss (4q, 5q, 16q,18q). Growth factors and cell adhesion molecules also may play a role (EGFR, c-erbB2 and src). However, a uniform molecular pathway has not been described.

Barrett's Mucosa
- Gastric fundus-type epithelium
- Junctional-type epithelium
- Specialized columnar epithelium – highest association with carcinoma

---

Current Concepts in General Surgery: A Resident Review, edited by William R. Wrightson.
©2006 Landes Bioscience.

Presentation

Persistent gastroesophageal reflux is the predominant indication that will alert a physician to the potential of malignant degeneration. The symptoms may progress, suggesting a more serious pathology.

1. Gastroesophageal reflux
2. Dysphagia
3. Weight loss

Diagnosis

Endoscopy

Diagnosis is established at endoscopy with histologic confirmation of columnar epithelial cells extending at least 3 cm into the esophagus above the anatomic GE junction. Current definitions now include any length of metaplasia to be of significance.

Management

The progress of Barrett's to severe dysplasia marks it as a premalignant lesion and current thought is to treat is as such with early resection (Table 10.1).

No or Mild Dysplasia – endoscopy with biopsy at 1-2 year intervals.
Moderate Dysplasia – endoscopy with biopsy at 6 month intervals
Severe Dysplasia – synonymous with carcinoma in situ and is an indication for resection.

Patients with chronic reflux should be aggressively treated medically. If this fails they should then undergo a fundoplication. Even after surgical intervention for reflux, Barrett's does not usually regress and requires constant surveillance. This is an important point in that performing fundoplication for reflux utilizes the future conduit for esophageal reconstruction and must be considered prior to surgery.

In recent reports, invasive carcinoma has been found in up to 50% of patients undergoing resection for severe dysplasia.

Maximum Review

- Accounts for 2-8% of esophageal cancers
- Presentation with GERD, dysphasia
- Diagnose with EGD and biopsy with mucosal changes 2 cm above the GE junction
- Surveillance of Barrett's
- Severe dysplasia is consistent with carcinoma in situ
- Early surgery can result in high cure rates

References

1.   Siewart JR, Stein HJ. Barrett's cancer: Indications, extent and result of surgical resection. Semin Surg Oncol 1997; 13:245-252.
2.   Cameron JL. Current Surgical Therapy. 6th ed. St Louis: Mosby 1998.

Motility Disorders

Motility disorders of the esophagus can be divided into three groups:

1. Achalasia
2. Diffuse esophageal spasm

Table 10.1. Staging for esophageal carcinoma

Primary Tumor (T)

TX	Primary tumor cannot be assessed (cytologically positive tumor not evident endoscopically or radiographically)
T0	No evidence of primary tumor (e.g., after treatment with radiation and chemotherapy)
Tis	Carcinoma in situ
T1	Tumor invades lamina propria or submucosa, but not beyond it
T2	Tumor invades muscularis propria
T3	Tumor invades adventitia
T4	Tumor invades adjacent structures (e.g., aorta, tracheobronchial tree, vertebral bodies, pericardium)

Regional Lymph Nodes (N)

NX	Regional nodes cannot be assessed
N0	No regional node metastasis
N1	Regional node metastasis

Distant Metastasis (M)

M1	Distant metastasis

Stage Grouping

Stage 0	Tis, N0, M0
Stage I	T1, N0, M0
Stage IIA	T2, N0, M0
	T3, N0, M0
Stage IIB	T1, N1, M0
	T2, N1, M0
Stage III	T3, N1, M0
	T4, Any N, M0
Stage IV	Any T, M1

10

  3. Nutcracker esophagus
  4. Hypertensive lower esophageal sphincter

Achalasia

Achalasia is characterized by esophageal aperstalisis, esophageal dilatation and failure of the lower esophageal sphincter to relax (Fig. 10.1).

Incidence

The incidence of achalasia in the US is 1/100,000; however this number may actually underestimate the true incidence.

Etiology

While the precise etiology of achalasia remains elusive, various theories have been proposed including infectious sources and heredity.

Pathophysiology

The development of achalasia is presumed secondary to neurodegeneration. Degeneration in the vagus nerve and Auerbach plexus have been shown histologically.

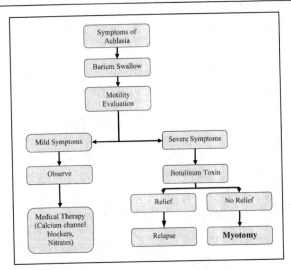

Figure 10.1. Management of achalasia.

This results in hypertension of the LES and increased esophageal pressure with loss of perstalsis.

Esophageal Spasm Syndromes

Presentation

Most patients will have chest pain or complain of dysphagia. The chest pain may in fact be difficult to distinguish from angina. Pain of cardiac origin must be ruled out prior to labeling this an esophageal problem.

Diffuse esophageal spasm results from hypertrophy of the mucsular layers and degeneration of the vagus. Manometry reveals multiple simultaneous or uncoordinated contractions. The lower esophageal sphincter (LES) will have short duration of relaxation and >20% of the distal esophagus will demonstrate simultaneous contractions.

Nutcracker esophagus is the most common of the primary esophageal motility disorders. It is associated with pressure amplitudes of >2 standard deviations from norm and distal esophageal pressures >180 mm Hg. Patients have normal peristaltic progression.

Hypertensive lower esophageal sphincter is associated with normal esophageal perstalsis but with delayed passage of food bolus through the gastroesophageal junction. The LES pressure is >26 mm Hg with relaxation pressures <8 mm Hg.

Diagnosis

Various studies are available to evaluate the esophagus. They include:

1. Barium esophagogram
2. Esophageal manometry
3. Video esophagogram
4. Endoscopy

Management

Esophageal myotomy is the surgical treatment when medical management fails. Long myotomy is performed through al muscle layers of the esophagus through the manometric defect and extended 2-3 cm beyond the GE junction. Following myotomy secondary reflux can be a problem and require surgical intervention. Surgical myotomy is associated with a low morbidity and gives superior long term results. However in practice, pneumatic dilitation is performed with a risk of rupture as high as 20%.

Modified Heller Myotomy

The myotomy is performed through all muscle layers. The myotomy is extended 2 cm beyond the GE junction and 4 cm beyond the defect proximally. A partial fundoplication can be added to prevent reflux disease.

Diverticula

There are three types of esophageal diverticula:
1. Zenkers (crycopharyngeal) diverticulum
2. Epiphrenic diverticulum
3. Traction diverticulum

Traction diverticula are typically midesophageal and associated with a mediastinal inflammatory process. Both epiphrenic and Zenkers are pulsion diverticula. Any patient discovered to have a esophageal diverticula should be studied for a motility disorder.

Zenkers is associated with dysphagia, regurgitation of undigested food, halitosis, aspiration. Diagnosis is made with a barium swallow. Endoscopy is not the test of choice in this case. The pathogenesis is the result of crycopharyngeal muscle dysfunction.

Management

Management of diverticula is the same as for other motility disorders with the application of a myotomy. This can be done under local or general anesthesia. Incision is made along the left SCM. The cervical esophagus is exposed by retraction of the SCM and carotid laterally and the trachea medially. The diverticulum is then isolated and can be tacked superiorly to the prevertebral fascia, or a diverticulectomy can be performed by firing a TA stapler across the diverticulum. The myotomy is done by dividing the muscle 2 cm proximal and 4 cm distal onto the esophagus.

Epiphrenic diverticula arise in the distal third of the esophagus. The management of the diverticulum is dependendant on the severity of the associated symptoms and is usually done with a myotomy to prevent suture line disruption of recurrence of the diverticulum.

References

1. In: Sabiston DC and Spencer FC, eds. Surgery of the chest. 6th ed. Philadelphia: W.B. Saunders Company, 1996.
2. Greenfield LJ, Mulholland M, Oldham KT et al. Surgery scientific principals and practice. 2nd ed. Philadelphia: Lippincott-Raven, 1997.
3. Sabiston DC, Lyerly HK. Textbook of surgery. 15th ed. Philadelphia: W.B.Saunders Co., 1997.

Esophageal Carcinoma

Anatomy

The esophagus is 24 cm in length. The distance from the incisors to the lower esophageal sphincter (LES) is 40 cm. Through the length of the esophagus there are three areas of narrowing:

1. Crycopharyngeal muscle (upper esophageal sphincter)
2. Aortic arch (24 cm)
3. LES (40 cm)

The entire esophagus lacks a serosa and has an extensive lymphatic plexus. This is a partial explanation as to occult spread of cancers beyond the gross specimen.

Surgical History

Billroth performed the first successful resection of a cervical esophageal carcinoma in 1871. The British surgeon, Turner performed the first transhiatal esophagectomy in 1933. Ivor Lewis popularized the right transthoracic approach to esophageal cancer in 1946.[1]

Background

The incidence of esophageal carcinoma in the US is 6/100,000 per year with a higher incidence found in black males (12/100,000). The incidence of esophageal cancer in Hunan China, Iran, and South Africa is in epidemic proportions. The natural history of esophageal has a dismal prognosis with a 0-7% 5 year survival.

10

Etiology

The exact etiology of esophageal carcinoma remains elusive. There has been a shift between squamous cell carcinoma to adenocarcimoma.

1. Tobacco
2. Alcohol
3. Hot beverages
4. Nitrosoamines

Pathophysiology

Esophageal carcinoma is generally divided into two principal histologic players: squamous cell carcinoma (SCC) and adenocarcinoma. Worldwide the most common histologic subtype is SCC although the incidence of adenocarcinoma is increasing.

Presentation

Esophageal carcinoma typically occurs in the sixth or seventh decades of life with a history of excessive use of tobacco and alcohol is common. Symptoms from esophageal carcinoma may be of insidious onset, beginning as nonspecific retrosternal discomfort or indigestion. As the tumor enlarges, symptoms progress with weight loss, odynophagia, chest pain, and occasionally hematemesis. Dysphagia is the presenting complaint in 80% to 90% of patients with esophageal carcinoma. Any adult who complains of progressive dysphagia warrants both a barium esophagogram and esophagoscopy to rule out carcinoma.

Table 10.2. Results of preoperative adjuvant therapy for esophageal cancer

	Patients	Preoperative Adjuvant Therapy	Resectable	Median Survival
Gignoux 1987	102	XRT	73%	11 mo
Roth 1988	17	Chemo	72%	10 mo
Orringer 1990	43	Chemo/XRT	91%	29 mo
Walsh 1996	58	Chwmo/XRT	90%	16 mo

Diagnosis

The combination of esophageal biopsy and brushings for cytologic evaluation establishes a diagnosis of carcinoma in 95% of patients with malignant strictures. The barium esophagogram, particularly using air contrast radiographic technique, has enabled the demonstration of lesions as small as 5 to 15 mm in early detection programs. Unfortunately, programs for early detection of esophageal carcinoma using mass screening of patients with barium esophagograms, flexible fiberoptic esophagoscopy, and cytology are not cost effective in Western cultures, where the incidence of this disease is relatively low.

Staging

Esophageal cancers are known for local invasion. The esophagus has a rich lymphatic plexus that allows for extension of a tumor beyond its macroscopic margins.

Work Up Includes

1. CT chest and abdomen
2. Bone scan (if symptomatic)
3. CT head (if symptomatic)
4. Pulmonary function studies (FEV_1>2 liters)—a FEV_1 <1.24 has an associated 40% risk of mortality within 4 years.
5. Cardiac evaluation (EF <40% have poor prognosis)

Cervical lesions are more common in women and if not extending into surrounding structures are amenable to surgery. Lesions at the thoracic inlet are typically advanced with extension into the great vessels. Only tumors in the thoracic esophagus that have not spread through the wall of the esophagus are resectable.

TNM Staging

Best outcomes are associated with early cancers (T1a-T1b Stage I). With 5 year survival approaching 90%. Most present at a later stage (II-IV) with an overall 5 year survival of 0-7% (Table 10.2).

Surgical Managemnt

Esophageal resection can be accomplished through various approaches. Currently the most accepted approaches include transhiatal resection and the Ivor-Lewis thoracoabdominal approach. Each has its advantages and disadvantages but there is no difference in long term and disease free survival between resection groups.

10

Thoracoabdominal Approach (Ivor-Lewis)

A midline incision is made and the stomach is mobilized. Preservation of the right gastric and gastroepiploic arteries is essential to utilize the stomach as an esophageal substitute. A pyloroplasty is performed to facilitate gastric emptying. A feeding jejunostomy may be placed. Once mobilization is complete the abdominal incision is closed and the patient is rotated to a left lateral decubitus position. Some surgeons advocate placing the patient in a modified supine/left decubitus position so both approaches can be done without moving the patient. A right thoracotomy through the 5th intercostal space is done with completion of the esophageal mobilization. The anastomosis is done with a single layer PDS or double layer chromic and silk.

Transhiatal Approach

This is similar to the Ivor-Lewis in terms of abdominal mobilization but without a thoracotomy. The esophagus is mobilized with blunt dissection through the esophageal hiatus. The anastomosis is performed in the neck with the esophageal substitute.

Disadvantages to the Ivor-Lewis include an intrathoracic anastomosis with a high mortality associated with leak and pulmonary complications associated with a thoracotomy. The transhiatal approach is done blindly with questionable margins. There is no difference in survival between the approaches. There are only three current randomized controlled studies that compare transhiatal to transthoracic resection. Survival was similar between groups with 29% and 26% for transthoracic and transhiatal resections respectively.[4]

Adjuvant Therapy

Use of chemotherapy combined with radiotherapy preoperatively (neoadjuvant therapy) has been shown to improve overall survival when compared to surgery alone. There are currently several randomized controlled studies underway to evaluate this further. The protocol includes 5-fluorouracin and cisplatinin with 4500 cGy radiotherapy over 3 weeks. There is a 3-6 week recovery period during which time WBC is followed with repeat CT scanning. The patient is restaged and surgery performed if possible. In the study conducted at the University of Michigan, 91% were resectable with a 24 month median survival compared to 14 month with surgery alone.[1] One significant randomized controlled study in 1996 showed a survival advantage in patients with adenocarcinoma treated with neoadjuvant therapy.[3]

References

1. Greenfield LJ, Mulholland M, Oldham KT et al. Surgery Scientific Principals and Practice. 2nd ed. Philadelphia: Lippincott-Raven, 1997.
2. Sabiston DC, Lyerly HK. Textbook of surgery. 15th ed. Philadelphia: W.B. Saunders Co., 1997.
3. Walsh TN, Noonan N, Hollywood D et al. A comparison of multi-modal therapy and surgery for esophageal adenocarcinoma. N Engl J Med 1996; 335:462-7.
4. Hulscher JBF, Tijssen JGP, Obertop H et al. Transthoracic versus transhiatal resection for carcinoma of the esophagus: A meta-analysis. Ann Thorac Surg 2001; 72: 306-313.

CHAPTER 11

Stomach and Duodenum
Steven M. Girard

Gastric Hormones

Gastrin

Gastrin is a peptide hormone that is produced by G-cells located primarily in the gastric antrum and, to a lesser, degree the duodenum. The most important stimulator of gastrin release is food intake. Specifically, amino acids and short polypeptides stimulate the release of gastrin. Additionally, G-cells are stimulated by acetylcholine via cholinergic enervation through the vagus nerve. Gastric distention results in inhibition of gastrin release via vagal enervation. Gastric pH is also a factor, which affects gastrin secretion. Low intraluminal pH inhibits gastrin release whereas a higher pH potentiates gastrin secretion. Somatostatin, conversely, inhibits gastrin release. Gastrin stimulates the secretion of hydrochloric acid by the parietal cell, which is present in the gastric fundus. In turn, hydrogen ion inhibits gastrin release via a negative feedback loop. Thus, in states of achlorhydria such as pernicious anemia, hypergastrinemia is observed. Finally, gastrin has a trophic effect on gastric mucosa.

Somatostatin

Somatostatin is another peptide hormone produced by endocrine cells of the pancreas and GI tract. Its production is stimulated through a number of mechanisms, not the least of which is gastric acidification. In short, somatostatin inhibits acid secretion by the parietal cell and gastrin release by G-cells.

Acid Secretion

There are numerous mediators which affect the secretion of acid by the parietal cell. In basic terms, the parietal cell is responsive to five main mediators which influence acid secretion. Three of these agents stimulate acid production while the other two have an inhibitory effect on acid production. Gastrin, histamine, and acetylcholine (via vagal enervation) all have a stimulating effect on acid production. Conversely, somatostatin and prostaglandin have an inhibitory effect on acid production.

References

1. Greenfield LJ, Mulholland M, Oldham KT et al. Surgery scientific principals and practice. 2nd ed. Philadelphia: Lippincott-Raven, 1997.
2. Sabiston DC, Lyerly HK. Textbook of Surgery. 15th ed. Philadelphia: W.B. Saunders Co., 1997.

Upper Gastrointestinal Hemorrhage

Etiology

Upper gastrointestinal hemorrhage is bleeding which occurs from a source located proximal to the ligament of Treitz. There are multiple sources of this type of bleeding. The most common cause of significant upper gastrointestinal bleeding is peptic ulcer disease. Generally, duodenal ulcer disease is a more frequent cause of bleeding than gastric ulcer (Table 11.1).

Presentation

History should provide a fairly accurate estimation of the source of upper GI bleeding. Information pertaining to previous episodes of bleeding is of obvious importance. The presence of concurrent diseases such as hepatic disease, alcoholism, peptic ulcer disease, and hematological disorders may provide additional clues. For example, a patient who reports chronic upper abdominal pain and ingestion of large amounts of NSAIDS probably has a gastric ulcer or erosive gastritis.

The manner in which the bleeding presents can provide clues to its source. For example, hematemesis of either bright red blood or "coffee grounds" suggests a bleeding source proximal to the ligament of Treitz. The presence of "coffee grounds" indicates that the hemoglobin has been in contact with gastric acid long enough to be converted to methemoglobin. Hematochezia, the passage of bright red blood per rectum, suggests a distal lower gastrointestinal source of bleeding. Alternatively, massive upper GI hemorrhage can present as hematochezia. Otherwise, upper GI bleeding typically produces melena. This indicates that the blood has been in the GI tract for a longer period of time.

Physical exam may also provide information that can suggest the likely cause of bleeding. Stigmata of portal hypertension (e.g., ascites, jaundice, caput medusa, palmar erythema, etc.) may point to esophageal varices as a likely cause of bleeding. Cachexia, an abdominal mass, and an enlarged Virchow's node may suggest underlying malignancy.

Diagnosis

It is important to identify the cause and location of upper GI bleeding. After the patient is resuscitated, upper endoscopy is performed to allow for identification of the bleeding site. In most cases this can be successfully accomplished. In addition, therapeutic measures such as heater coagulation, epinephrine injection, and variceal banding can be performed at the same setting after diagnosis is confirmed. Before

Table 11.1. Causes of upper gastrointestinal bleeding

Source	Frequency
Peptic ulcer disease	45%
Gastritis	30%
Esophageal varices	10%
Mallory-Weiss tear	10%
Gastric carcinoma	5%

upper endoscopy can be performed, placement of a nasogastric tube can help distinguish upper GI bleeding from lower. Basically, a bleeding site proximal to the ligament of Treitz is suggested when blood is aspirated from a properly placed nasogastric tube. Angiography and tagged red blood cell scan can also help to identify the site of bleeding but are not commonly needed in cases of upper gastrointestinal hemorrhage.

Treatment

Immediate resuscitation is the first treatment priority with upper GI bleeding. The volume of blood loss can be deduced by physical findings. Tachycardia and narrowing of the pulse pressure are sensitive early indicators of blood loss. Orthostatic hypotension and subtle mental status changes may be present. Hypotension and oliguria are late signs which indicate a large volume of blood loss. The patient's physiologic response to volume infusion is important. Persistent tachycardia despite volume resuscitation is an ominous sign of significant ongoing hemorrhage.

Nasogastric tube placement and gastric lavage are performed to remove pooled blood, which reduces fibrinolysis at bleeding sites. Lavage and evacuation of the stomach also prevent gastric distention, which predisposes the patient to vomiting and aspiration. Gastric distention also stimulates gastrin release. Correction of coagulopathy is essential when present. Blood transfusion is performed when necessary.

Gastric pH should be mainained at >5.0 with antacids, H2 blockers, or proton pump inhibitors. This usually does not stop ongoing bleeding but is necessary to limit progression of disease and allow for healing. Endoscopic electrocautery or epinephrine injection may be attempted with bleeding ulcers or gastritis. Sclerotherapy or endoscopic banding may be attempted with variceal bleeding. Angiographic embolization does not work well because of the rich submucosal vascular plexus of the stomach. Selective infusion of vasopressin into the left gastric artery may lead to a temporary response.

Surgical Management

Surgical treatment is indicated in cases where medical treatment fails to lead to cessation of bleeding. Fortunately, most upper GI hemorrhage will cease with conservative measures. Certain patient characteristics may lead one to consider surgery earlier. For instance, large ulcers with a visible vessel in the base are likely to experience significant rebleeding and probably should be operated on. In addition, patients with significant cardiac disease in whom bleeding would be poorly tolerated may be considered earlier for surgery. Other considerations for early surgery would be an existing contraindication to transfusion (e.g., Jehovah's witness), a difficult cross-match, or flow-dependent cerebrovascular disease.

Selected Reading

1. Nyhus LM, Baker RJ, Fischer JE. Mastery of surgery. 3rd ed. Boston: Little, Brown and Co., 1997.
2. Greenfield LJ, Mulholland M, Oldham KT et al. Surgery scientific principals and practice. 2nd ed. Philadelphia: Lippincott-Raven, 1997.
3. Sabiston DC, Lyerly HK. Textbook of surgery. 15th ed. Philadelphia: W.B. Saunders Co., 1997.

11

Helicobacter pylori

Background

Helicobacter pylori is a spiral bacterium which specifically colonizes gastric mucosa. It is able to survive in a near neutral pH through the production of urease. *H. pylori* infection has been correlated with the formation of peptic ulcers. This association was first suggested in 1984. Several lines of evidence link *H. pylori* to peptic ulceration.

- First, eradication of *H. pylori*, without the suppression of acid, leads to ulcer healing rates that are similar to those of acid suppression therapy alone.
- Secondly, relapse of duodenal ulceration after antimicrobial treatment is preceded by reinfection of gastric mucosa.
- Additionally, the prevalence of *H. pylori* in patients with duodenal ulceration is near 100%. Patients with gastric ulcers have a prevalence of *H. pylori* in the range of 60-80%. Comparatively, Helicobacter colonizes approximately 20% of those without peptic ulcer disease.

Consequently, the evaluation for and the treatment of Helicobacter infection is an integral part of the treatment of peptic ulcer disease.

Diagnosis

The diagnosis can be made with a biopsy of the gastric mucosa. This is subjected to histologic evaluation for the organism. The presence of urease producing organisms also suggests *H. pylori*. The urea breath test is another option. Current ELISA tests are available. The sensitivity and specificity of these tests approach 90%.

Management

The antimicrobial treatment of *Helicobacter pylori* is an important component of the treatment of peptic ulceration. *H. pylori* is associated with duodenal ulceration in nearly 100% of cases and with gastric ulceration in about 60% of cases. Helicobacter infection should be considered when patients experience ulcer recurrence on maintenance medical therapy or when peptic ulcers fail to heal.

Multiple medical regimens exist consisting of some combination of clarithromycin, metronidazole, bismuth, and/or tetracycline. Triple therapy is becoming more popular with use of a PPI and two antibiotics (i.e., omeprazole, clarithromycin and metrinidazole). Regardless of the exact regimen used, a 10-14 day course leads to eradication of *H. pylori* in approximately 90% of cases. This results in a decreased rate of recurrence in ulcer disease (6-25% compared to 70-80% without eradication).

References

1. Greenfield LJ, Mulholland M, Oldham KT et al. Surgery scientific principals and practice. 2nd ed. Philadelphia: Lippincott-Raven, 1997.
2. Sabiston DC, Lyerly HK. Textbook of surgery. 15th ed. Philadelphia: W.B. Saunders Co., 1997.

Peptic Ulcer Disease and Complications of Management

Background

Peptic ulcer disease remains a significant health problem in the US. There are over 300,000 cases annually with over 4 million people on some form of antiulcer therapy. They occur more commonly in men with duodenal ulcers two times more common that gastric ulcers.

Pathophysiology

The pathophysiology of peptic ulcers, whether gastric or duodenal, is multifactorial. Ulcers develop depending on the balance of injurious factors and host defense mechanisms. Environmental stressors are numerous. Acid secretion is involved in the final common pathway of ulcer formation. In regards to duodenal ulceration, mucosal exposure of acid is a necessary component of their pathogenesis. In general, patients with duodenal ulcers have an increased capacity for acid secretion. Maximal acid output in normal patients is usually around 20 mEq/hour but patients with duodenal ulcers may exceed 40 mEq/hr. There is, however, variability in this finding to the extent that some patients with duodenal ulcers have maximal acid outputs that are considered in the normal range. Additionally, patients with duodenal ulcer have increased basal acid outputs with normal basal gastrin levels. The role that acid secretion plays with the formation of gastric ulcers is less clear. Hypersecretion of acid is not associated with all subtypes of gastric ulcers. Gastric acid does, however, play a role in the chain of events that leads to ulceration.

Helicobacter pylori

11

Helicobacter pylori is a spiral bacterium which specifically colonizes gastric mucosa. It is able to survive in a near neutral pH through the production of urease. *H. pylori* infection has been correlated with the formation of peptic ulcers. Several lines of evidence link *H. pylori* to peptic ulceration.

- First, eradication of *H. pylori*, without the suppression of acid, leads to ulcer healing rates that are similar to those of acid suppression therapy alone.
- Secondly, relapse of duodenal ulceration after antimicrobial treatment is preceded by reinfection of gastric mucosa.
- Additionally, the prevalence of *H. pylori* in patients with duodenal ulceration is near 100%. Patients with gastric ulcers have a prevalence of *H. pylori* in the range of 60-80%. Comparatively, Helicobacter colonizes approximately 20% of those without peptic ulcer disease.

Consequently, the evaluation for and the treatment of Helicobacter infection is an integral part of the treatment of peptic ulcer disease.

There are several other environmental factors which are associated with the development of peptic ulcers. Cigarette smoking has been associated with peptic ulcer disease in so far as it impairs healing, decreases the effectiveness of therapy, increases recurrence rates, and increases the likelihood of complications. Ethanol also has injurious effects on gastroduodenal mucosa. NSAID use has also shown ulcerogenic effects. Ulcer formation is promoted through the systemic effects of cyclooxygenase inhibition and a direct toxic effect on mucosal cells. The use of NSAIDS is more strongly associated with gastric ulcers.

Finally, altered host defense mechanisms play a role in the pathogenesis of peptic ulcer formation. Surface epithelial cells produce bicarbonate and mucus. The generation of these substances allows for the creation of a pH gradient at the luminal interface such that there is a near neutral environment at the mucosal surface. Additionally, prostaglandins PGE2 and PGI2 have a protective effect on the mucosa through inhibition of acid secretion by the parietal cell. Patients with peptic ulcer disease have been shown to have decreased bicarbonate secretion and decreased production of mucosal prostaglandins. Motility disorders can also promote ulcerogenesis presumably through a mechanism of impaired clearance of noxious substances.

Presentation

The chief presenting complaint of patients with peptic ulcer disease is abdominal pain. The pain is usually referred to the epigastrium or left upper quadrant and described as gnawing, burning, or stabbing. With duodenal ulcers, relief is frequently experienced with food intake. In contradistinction, patients with gastric ulcers will frequently have exacerbation of pain with food intake. Nausea and vomiting may be present but are not major features of this disease.

Duodenal ulcers are found in the first portion of the duodenum in about 95% of cases. The presence of ulcers in more distal areas of the duodenum is atypical and should arouse clinical suspicion of an underlying gastrinoma. Gastric ulcers most commonly occur on the lesser curvature. They are located just above the incisura angularis on the lesser curvature in about 60% of cases. An additional 15-25% are located distal to the incisura on the lesser curvature and another 10% are located high on the lesser curvature. Only about 5% are found on the greater curvature of the stomach.

Diagnosis

The diagnosis of peptic ulcer disease is best verified by upper endoscopy. Esophagogastroduodenoscopy is greater than 95% sensitive and near 100% specific in identifying peptic ulcers of both the duodenum and stomach. There are several endoscopic features, additionally, of gastric ulcers that can help to distinguish between peptic ulceration and malignant ulceration. Benign ulcers typically appear as round lesions with slightly raised, smooth borders. The surrounding mucosal folds are symmetric and taper evenly toward the edge of the ulcer. In addition, benign ulcers often have a smooth base covered with a fibrous layer. Although these characteristics may suggest a benign ulcer the only way to truly distinguish is with multiple biopsies at the ulcer edge. Malignancy must be ruled out with biopsy upon the discovery of a gastric ulcer.

Subtypes of Gastric Ulcers

Gastric ulcers are classified into one of four different subtypes:

- **Type I ulcers** occur along the lesser curvature, in the body of the stomach, just above the incisura angularis. These ulcers account for about 40-50% of gastric ulcers and as such are the most common type of gastric ulcer. Gastric acid output is within the normal range. Surgical therapy consists of a distal gastrectomy with gastrojejunostomy. Vagotomy is not necessary since these ulcers are not associated with acid hypersecretion. In fact, the addition of a vagotomy does not decrease the ulcer recurrence rate, which is about 3%.

- **Type II ulcers** also occur along the lesser curvature, in the body of the stomach, and again are found just above the incisura. These ulcers, however, are associated with the simultaneous presence of a duodenal ulcer. They account for about 25% of gastric ulcers. Type II ulcers are associated with gastric acid hypersecretion. Surgical therapy generally consists of a truncal vagotomy with either antrectomy or pyloroplasty.
- **Type III ulcers** are prepyloric ulcers. They account for about 25% of gastric ulcers. These ulcers are also associated with gastric acid hypersecretion. Surgical therapy consists of truncal vagotomy with either antrectomy or pyloroplasty. Parietal cell vagotomy is associated with higher rates of ulcer recurrence when used for the treatment of type III ulcers.
- **Type IV ulcers** occur high on the lesser curvature, near the GE junction. They account for less than 10% of gastric ulcers. Type IV ulcers are associated with normal levels of gastric acid secretion. Surgical therapy is more complicated than with the other types of gastric ulcers and depends on the proximity of the ulcer to the GE junction. Esophagogastrectomy may be necessary for ulcers too close to the distal esophagus to allow for preservation of the GE junction.

Medical Therapy

H2 receptor antagonists competitively bind histamine receptors on the surface of the gastric parietal cell. Several different H2 blockers are clinically available. However, there is no significant difference among these different H2 blockers in regards to their efficacy in healing ulcers. About 70% of patients are ulcer-free after 4 weeks of treatment. Up to 90% of patients will be ulcer-free at the end of 8 weeks of treatment.

11

Proton pump inhibitors bind to membrane-bound H+/K+/ATPase on parietal cells and in doing so block the intraluminal secretion of hydrogen ion. Direct comparisons between omeprazole and H2 blockers have demonstrated that omeprazole is superior in regards to pain relief and ulcer healing. About 80% of patients are ulcer-free after 2 weeks of treatment. Up to 95% are ulcer-free at the end of 4 weeks of treatment.

Sucralfate is activated at a pH of <3.5 and polymerizes to form an insoluble gel which binds to proteins on injured mucosa. In doing so it forms a barrier between injured mucosa and luminal acid to prevent further acid-induced injury. Sucralfate also binds free bile salts and pepsin, reducing their ability to cause mucosal damage. Additionally, it stimulates mucosal production of mucus, bicarbonate, and prostaglandins. As it does not influence acid secretion, sucralfate does not promote bacterial overgrowth within the stomach. Acid-reducing therapy with proton pump inhibitors or H2 receptor antagonists raise intraluminal gastric pH and consequently decrease the efficacy of sucralfate when these medications are used in combination. Sucralfate displays ulcer-healing efficacy similar to that of H2 blockers.

Antacids work by neutralizing gastric acid. They have ulcer-healing efficacy comparable to H2 blockers. However, effective treatment with antacids requires frequent daily dosing which negatively influences patient compliance.

The antimicrobial treatment of *Helicobacter pylori* is an important component of the treatment of peptic ulceration. As previously mentioned, *H. pylori* is associated with duodenal ulceration in nearly 100% of cases and with gastric ulceration in about 60% of cases. Helicobacter infection should be considered when patients experience ulcer recurrence on maintenance medical therapy or when peptic ulcers fail to heal. Multiple medical regimens exist consisting of some combination of clarithromycin, metronidazole, bismuth, and/or tetracycline. Triple therapy is becoming more popular with use of a PPI and two antibiotics (i.e., omeprazole, clarithromycin and metrinidazole). Regardless of the exact regimen used, a 10-14 day course leads to eradication of *H. pylori* in approximately 90% of cases.

Surgical Therapy

In simplest terms, there are three basic operations for uncomplicated peptic ulceration: parietal cell vagotomy, truncal vagotomy with pyloroplasty, and truncal vagotomy with antrectomy. With the advent of H2 receptor antagonists and, more recently, proton pump inhibitors the surgical treatment of peptic ulcer disease has become less common. Surgical treatment is indicated when ulcers fail to heal after 3 months of appropriate medical therapy or when ulcers recur on maintenance therapy. Surgery is also indicated when malignancy cannot be excluded and when complications of disease occur. Complications of peptic ulcer disease include perforation, hemorrhage, and obstruction.

Parietal cell vagotomy selectively inhibits vagal stimulation of parietal cells and smooth muscle cells of the gastric fundus. It spares the vagal enervation to the antrum, pylorus, small bowel, biliary tract, and pancreas. Acid secretion is diminished by the interruption of vagal stimuli to parietal cells. Specifically, basal acid secretion is decreased by about 80% and maximal acid secretion is decreased by about 70%. There is some rebounding of both basal and maximal acid secretion over time but neither rebound to preoperative levels. Vagal denervation of the gastric fundus inhibits receptive relaxation of the fundus. As a result, gastric emptying of liquids is increased. As the antrum and pylorus are spared, there is no effect on the emptying of solids.

Truncal vagotomy has similar efficacy in regards to the reduction of acid secretion. As with parietal cell vagotomy, truncal vagotomy decreases receptive relaxation of the gastric fundus increasing the emptying of liquids. Truncal vagotomy additionally inhibits antral and pyloric motility which results in poor emptying of solids. Pyloroplasty is included to overcome the effect of diminished gastric emptying. It effectively provides a wider gastric outflow tract so that the emptying of solids is increased.

Antrectomy removes the bulk of gastrin producing cells and effectively reduces basal gastrin levels by 50% and postprandial gastrin levels by 67%. Reconstruction of the upper GI tract is via gastroduodenostomy (Billroth I) or loop gastrojejunostomy (Billroth II). Truncal vagotomy and antrectomy results in the reduction of basal and maximal acid secretion by about 85%. Inhibited fundic receptive relaxation again results in the increased emptying of liquids. The emptying of solids is decreased. Both forms of reconstruction are similar in regards to operative mortality, morbidity, and rates of recurrence. Comparison of these three procedures is as follows: (Table 11.5).

Complications of Peptic Ulcer Disease

Hemorrhage is the leading cause of death associated with peptic ulcer disease. Lifetime risk of bleeding for those without treatment is approximately 35%. Spontaneous, and often temporary, cessation of bleeding occurs in about 70% of patients. EGD is appropriate in the acute setting. This is successful in identifying the cause of bleeding in the majority of cases. Sometimes bleeding can be controlled with thermal coagulation via a heater probe. Stigmata of recent bleeding include the presence of an adherent clot or the presence of a visible vessel in the ulcer base. In general, approximately 30% of those with stigmata of recent bleeding will rebleed and a large proportion of these will require emergent operation. Operative strategy is that of duodenotomy with oversewing of the bleeder combined with a definitive acid-reducing procedure. Operative intervention is necessary when there is ongoing massive hemorrhage leading to shock or cardiovascular instability, prolonged blood loss requiring transfusion, or recurrent bleeding despite medical or endoscopic therapy.

Perforation. The lifetime risk of perforation in untreated peptic ulcer disease is approximately 10%. Perforation is unusual if ulcer healing has been accomplished with medical therapy. Clinically, patients demonstrate sudden onset of severe diffuse abdominal pain, which quickly reaches peak intensity and remains constant. Free air is often noted on plain, upright chest X-ray. However, free air may not be seen in as many as 20% of cases. The site of perforation is closed with an omental patch (Graham patch) and irrigation of the peritoneal cavity. Simple omental patching is associated with an 80% chance of recurrent ulceration and a 10% rate of repeat perforation. This is not ideal and often definitive antiulcer therapy is necessary eventually. The timing of a definitive procedure is dictated by the physiologic status of the patient and is not performed if the patient has preoperative shock, life-threatening coexistent medical illness, or if there is a significant delay in diagnosis. In these situations, an omental patch and peritoneal wash-out is performed and the patient is brought back to the operating room at another time when he or she has had an opportunity to recover.

Obstruction can occur acutely as a result of edema and inflammation or chronically as a result of scarring. Obstruction is associated with ulcers of the pyloric channel or bulb of the duodenum. Acute obstruction can, in general, be treated conservatively with rehydration, correction of electrolyte abnormalities, and nasogastric tube decompression. Chronic obstruction occurs after repeated ulceration followed by healing. With untreated ulcer disease the lifetime risk of obstruction is approximately 10%. EGD is indicated to confirm the diagnosis and to exclude malignancy. Approximately 85% of cases of chronic obstruction are amenable to hydrostatic balloon dilatation. About 80% of these patients will experience immediate relief. About 40% will still be unobstructed at 3 months. Repeated dilation is sometimes necessary. Operative management consists of relief of the obstruction with additional definitive antiulcer surgery.

References

1. Nyhus LM, Baker RJ, Fischer JE. Mastery of Surgery. 3rd ed. Boston: Little, Brown and Co., 1997.
2. Greenfield LJ, Mulholland M, Oldham KT et al. Surgery Scientific Principals and Practice. 2nd ed. Philadelphia: Lippincott-Raven, 1997.
3. Sabiston DC, Lyerly HK. Textbook of Surgery. 15th ed. Philadelphia: W.B. Saunders Co., 1997.

Postgastrectomy Syndromes

Dumping syndrome is related to unchecked entry of ingested food into the proximal small bowel. Dumping syndrome involves both vasomotor and abdominal complaints. Vasomotor symptoms include dizziness, sweating, palpitations, and sometimes syncope. Abdominal complaints include epigastric pain, nausea, and borborygmi. Dumping can be divided into early and late symptoms. Early dumping occurs immediately after ingestion of a meal. Late dumping occurs 1-3 hours after a meal and sometimes can involve reactive hypoglycemia. In general, late dumping syndrome involves more vasomotor symptoms and fewer abdominal symptoms. The incidence of dumping increases as more aggressive operations are performed (i.e., there is a greater incidence of dumping with antrectomy than with pyloroplasty). Fortunately, long-term debilitating dumping persists in about 1-2% of patients. Subcutaneous octreotide (50-100 mcg) before meals results in symptomatic improvement in approximately 70-90% of patients with either early or late syndromes. In addition, dietary modification can lead to improvement of symptoms. These modifications include eating smaller meals, which are low in carbohydrates, avoidance of liquids during meals, and intake of room temperature hypotonic liquids.

Alkaline reflux gastritis presents with symptoms of postprandial epigastric pain and nausea. Diagnosis is confirmed objectively by gastric fluid analysis demonstrating bile and histiologic evidence of gastritis. This occurs transiently in approximately 10-20% of patients undergoing vagotomy and either pyloroplasty or antrectomy. Persistent symptoms occur in 1-2% of patients. Treatment with antisecretory drugs, bile acid chelators, and dietary modification does not lead to consistent control of symptoms. Conversion to a Roux-en-Y gastrojejunostomy with a 50-60 cm limb may be necessary for intractable symptoms in order to divert the biliary stream. This effectively eliminates biliary emesis in all patients. Up to 30% of patients may experience persistent pain despite conversion and about 20% will see a delay in gastric emptying.

Afferent loop obstruction is a condition characterized by intermittent epigastric or right upper quadrant pain, which is relieved by bilious vomiting. The vomiting is often projectile and consists of bile without food. The acute postoperative form of this condition predisposes the patient to duodenal stump "blow-out". The chronic form may result in marked bacterial overgrowth of the chronically obstructed loop. Diagnosis may be made by radionuclide biliary scanning which demonstrates a dilated afferent limb. CT scan may lead to the diagnosis. Barium upper GI can also sometimes suggest the diagnosis by demonstrating a trickle of contrast into the dilated afferent loop. Upper endoscopy may rule out the diagnosis by demonstrating a widely patent afferent portion of the anastomosis. Preferred therapy is surgical and consists of conversion to a Roux-en-Y gastrojejunostomy with a 50-60 cm Roux limb. Revision of the loop gastrojejunostomy is also an option.

Retained gastric antrum is a syndrome that occurs in association with a Billroth II reconstruction following antrectomy. It occurs when a portion of antrum is left adjacent to the duodenal stump. This antral tissue contains gastrin-producing cells, which are isolated from the negative-feedback effects of gastric acid. Hypergastrinemia results with the complication of recurrent ulceration.

Postvagotomy diarrhea may occur in up to 20% of patients after vagotomy and either pyloroplasty or antrectomy. However, only 1-2% will have persistent incapacitating symptoms. The cause of diarrhea is unclear but may be related to alterations of bowel motility combined with increased excretion of fecal fat. Conservative management consists of dietary modification. Improvement may be observed with more frequent, smaller meals, ingestion of meals with less liquid content, and an increase in dietary fiber. Antidiarrheal medications may also be employed with varying efficacy. In refractory cases, surgery may be attempted. An effective strategy is the interposition of a 10 cm antiperistaltic segment of jejunum approximately 100 cm past the ligament of Treitz. It is rare for the syndrome to proceed to the point where surgical options are entertained.

The Roux syndrome is a syndrome of postoperative gastroparesis following vagotomy and antrectomy with a Roux-en-Y gastrojejunostomy. The early syndrome is characterized by prolonged hospitalization following Roux-en-Y gastrojejunostomy with persistent vomiting. Upper GI will occasionally suggest complete gastric outlet obstruction but endoscopy will demonstrate a widely patent anastomosis. Conservative measures consisting of intravenous alimentation, nasogastric tube decompression, and possible prokinetic agents often will lead to improvement over time. The late syndrome is more subtle and characterized by episodic vomiting, pain, and epigastric fullness often with normal endoscopic and upper GI examinations. It has been suggested that this syndrome is caused by the creation of a Roux limb, which has bypassed the duodenal pacemaker and, thus, alters the propagation of myoelectric activity. Complete vagal denervation is necessary in the pathogenesis of this syndrome. Prokinetics should be initiated prior to consideration of corrective surgery. When medical treatment fails, surgery may be attempted with an operative strategy consisting of partial gastrectomy with Roux reconstruction.

References

1. Nyhus LM, Baker RJ, Fischer JE. Mastery of surgery. 3rd ed. Boston: Little, Brown and Co., 1997.
2. Greenfield LJ, Mulholland M, Oldham KT et al. Surgery scientific principals and practice. 2nd ed. Philadelphia: Lippincott-Raven, 1997.
3. Sabiston DC, Lyerly HK. Textbook of surgery. 15th ed. Philadelphia: WB Saunders Co., 1997.

11

Gastric Adenocarcinoma and Lymphoma

Background

The incidence of gastric adenocarcinoma has been on the decline over the last century. In 1930 the incidence was about 40 cases per 100,000 population. By 1980 the incidence had dropped to about 10 cases per 100,000. Over the last 20 years, however, this rate has been fairly constant. The highest age-adjusted death rates for gastric adenocarcinoma occur in Japan where it accounts for 50% of cancer-related deaths in men and 40% in women.

Etiology

The etiology of gastric adenocarcinoma is believed to be largely environmental. Migration from an area of high-risk to an area of low-risk results in a decreased probability of developing gastric adenocarcinoma. Several environmental factors have been associated with the development of gastric adenocarcinoma. Ingested nitrates and nitrosamines have been shown to promote carcinogenesis in animal models. In addition, *Helicobacter pylori* may also have a role in carcinogenesis.

Premalignant Conditions

As with colorectal adenocarcinoma, certain gastric polyps are considered premalignant. Most gastric polyps are asymptomatic. When they are symptomatic the most common complaint is of vague upper abdominal pain. Hemorrhage is infrequent and complications are rare.

Hyperplastic polyps are fairly common, occurring in about 0.5 to 1% of the general population. Hyperplastic polyps account for about 70-80% of all gastric polyps. They represent an overgrowth of normal gastric mucosa. Atypia is rare and they do not have malignant potential.

Adenomatous polyps are associated with an increased incidence of malignant disease. Mucosal atypia is more common. Dysplasia and carcinoma in situ may develop over time. Larger polyps and numerous polyps are associated with a higher risk of developing malignant disease. Endoscopic removal is indicated and considered sufficient for pedunculated lesions if the polyp is completely removed and no invasive cancer is found on pathologic examination. Endoscopic surveillance is indicated in the patient who demonstrates adenomatous polyps. Approximately 50% of patients with familial adenomatous polyposis have gastric polyps.

Some forms of chronic gastritis are also associated with gastric adenocarcinoma. Patients with type A chronic gastritis associated with pernicious anemia have a risk of carcinoma about twice that of the general population. Fundic mucosal atrophy, hypochlorhydria, loss of parietal and chief cells, and hypergastrinemia characterize this form of chronic gastritis. Routine surveillance of these patients is not generally performed but any new symptoms warrant aggressive work-up. Conversely, type B (antral) gastritis is not considered a risk factor for the development of carcinoma.

Presentation

Symptoms of gastric adenocarcinoma are generally vague and nonspecific. Epigastric pain is the most common symptom and is present in about 70% of patients. Pain is described as constant, nonradiating, and unrelieved by food. Temporary relief with antacids and antisecretory medications is sometimes observed. Dysphagia may be associated with more proximal gastric lesions. Frank upper GI bleeding occurs in about 10% of patients but up to one-third will be guiac positive. Perforation is an uncommon complication. The presence of cachexia, a palpable abdominal mass, hepatomegaly, or enlarged supraclavicular (Virchow's) nodes are ominous signs and suggest advanced disease.

The diagnosis of gastric carcinoma is generally made with upper endoscopy. Carcinomas may appear as polypoid, plaque-like, or ulcerative lesions. There are several features of malignant ulcers that differ from benign lesions. Malignant ulcers often have irregular, heaped-up borders. The ulcer base is more often irregular and necrotic. An underlying mass may be suggested. In addition, the surrounding gastric folds are often irregular and asymmetric in malignant ulcers. The diagnosis of gastric carcinoma is made in greater than 95% of cases if multiple biopsies are taken from the raised border of the ulcer. Mass population screening, while of value in Japan, has not been strongly advocated in the U.S. because of the lower incidence of the disease in this country.

Pathology

There are several pathologic variants of gastric adenocarcinoma. In the intestinal form, the malignant cells form gland-like structures. This type is more frequently associated with gastric mucosal atrophy, chronic gastritis, intestinal metaplasia, and dysplasia. This is the most common pathological variant.

The diffuse variant of gastric adenocarcinoma lacks gland formation. It is characterized, rather, by sheets of loosely adherent cells. This type tends to occur in younger patients and accounts for a higher proportion of cases that occur in low-risk areas. It is less common and associated with a poorer prognosis than the intestinal type.

Gastric carcinomas occur with equal frequency in the proximal and distal stomach. About 40% occur in the proximal stomach, 40% in the distal stomach, and about 20% have diffuse involvement. Distal tumors tend to have a better prognosis. Approximately 15% of gastric cancers will have nodal metastasis at the time of diagnosis.

TNM Classification

The TNM classification system stages tumors based on the depth of tumor invasion, nodal involvement, and the presence or absence of distant metastasis. A summary of this classification is as follows (Tables 11.2, 11.3 and 11.4):

Management

Surgical resection affords the only hope for cure. However, an advanced stage of disease at the time of diagnosis precludes this possibility in many patients. Carcinomas of the distal stomach are often treated with subtotal gastrectomy. More proximal lesions may require total gastrectomy or even esophagogastrectomy for adequate resection. These tumors often demonstrate extensive intramural spread. For this reason, wide margins of resection around the tumor are required. Retrospective studies have suggested that margins of 6 cm around the gross tumor are required to minimize the rate of local recurrence. As would be expected, microscopic disease at the

11

Table 11.2. Staging for gastric cancer

Tumor		
	T1	tumor confined to mucosa
	T2	involves mucosa and submucosa, extends to but does not penetrate serosa
	T3	penetrates serosa with or without invasion of adjacent structures
	T4	diffuse involvement on gastric wall (linitis plastica)
Nodes		
	N0	no nodal metastasis
	N1	metastasis to perigastric lymph nodes in immediate vicinity of tumor
	N2	metastasis to lymph nodes distant from primary tumor or along both curvatures of the stomach
Metastasis		
	M0	no distant metastasis
	M1	metastasis beyond regional lymph nodes

Table 11.3. Stage groupings for gastric carcinoma

Stage	Grouping
Stage I	T1, N0, M0
Stage II	T2-3, N0, M0
Stage III	T1-3, N1-3, M0
Stage IV	Tumor unresectable or metastatic

Table 11.4. Survival rates for gastric carcinoma

Stage	Survival
Stage I	approaching 100% 5-year survival
Stage II	50% 5-year survival, approx. 70% 1-year survival
Stage III	10% 5-year survival, approx. 60% 1-year survival
Stage IV	< 5% 5-year survival, approx. 20% 1-year survival

Table 11.5. Complications of anti-ulcer surgery

	Parietal Cell Vagotomy	Vagotomy and Pyloroplasty	Vagotomy and Antrectomy
Mortality	< 0.5%	0.5-1.0%	1-2%
Ulcer recurrence	10%	5-10%	1-2%
Dumping	< 5%	10%	10-15%
Diarrhea	< 5%	25%	20%

11

margin of resection is associated with a high rate of local recurrence and a poorer prognosis. Extended lymphadenectomy is commonly practiced in Japan. The value of this, however, is debated.

Palliative surgery is commonly entertained in cases of advanced disease. Surgical palliation often takes the form of gastric resection. While resection does not extend survival over that seen with bypass of obstructing lesions, it does provide improved symptomatic relief. Mean survival after palliative resection is approximately 9 months. Endoscopic laser fulguration is also sometimes an option for palliation.

Chemotherapy for gastric adenocarcinoma is of limited utility. Single-agent treatment regimens with 5-FU, mitomycin C, or doxorubicin have led to partial responses in about 25% of cases. Better response rates have been seen with multi-agent regimens, but survival rates have not been significantly influenced. Adjuvant chemotherapy after potentially curative surgery has no proven benefit. Radiotherapy has shown only modest benefits.

Gastric Lymphoma

Background

Nonhodgkin's lymphoma accounts for about 5% of all gastric malignancies. While gastric lymphoma is relatively uncommon compared to adenocarcinoma, the stomach is the most common site of extranodal lymphoma. The peak incidence of this disease is in the 6th and 7th decades.

Presentation

When gastric lymphoma is symptomatic, its symptoms are similar to those of gastric adenocarcinoma. Gross bleeding is uncommon but occult hemorrhage and anemia may be present in up to 25% of cases. Diagnosis is generally made via upper endoscopy and biopsy.

When the diagnosis of gastric lymphoma is made, evidence of systemic disease should be sought. A careful examination of lymph nodes is made. In addition, CT scans of the chest and abdomen are generally acquired. Bone marrow biopsy is also sometimes performed.

Staging of primary gastric lymphoma is as follows:
- Stage I: disease confined to the stomach
- Stage II: disease involving the perigastric lymph nodes
- Stage III: disease spread to lymph nodes beyond perigastric nodes
- Stage IV: disease spread to other solid organs

Managemant

Approximately 75% of patients with primary gastric lymphoma will have resectable disease at the time of diagnosis. Patients with stage I disease have disease which is generally curable with resection. Surgical therapy includes total gastrectomy with intraoperative staging. Data suggests that adjuvant radiotherapy improves locoregional disease control. In patients with stage II disease, 30% of those who undergo curative resection and adjuvant radiotherapy will have disease recurrence outside the treatment field. Thus, stage II disease is considered systemic disease and adjuvant chemotherapy is included in the treatment regimen.

In patients with stage III and IV disease, treatment primarily consists of chemoradiation. Surgery is performed in those patients who suffer complications of disease (i.e., bleeding, obstruction, or perforation) or who have persistent gastric disease following chemoradiation. If patients present with these complications, surgery is performed first followed by chemoradiation. Chemotherapy consists of cyclophosphamide, doxorubicin, vincristine, and prednisone (CHOP).

Stage I disease is associated with a 5-year survival rate of approximately 90%. Stage II is associated with a 75% 5-year survival. Stage III disease is associated with a 50% 5-year survival while stage IV disease is associated with a 5-year survival rate of about 15%.

11

References

1. Nyhus LM, Baker RJ, Fischer JE. Mastery of Surgery. 3rd ed. Boston: Little Brown and Co., 1997.
2. Greenfield LJ, Mulholland M, Oldham KT et al. Surgery scientific principals and practice. 2nd ed. Philadelphia: Lippincott-Raven, 1997.
3. Sabiston DC, Lyerly HK Textbook of surgery. 15th ed. Philadelphia: W.B. Saunders Co., 1997.

Gastrinoma and the Zollinger-Ellison Syndrome

Background

A gastrinoma is a neoplasm which liberates the active hormone gastrin. Increased circulating levels of gastrin stimulate parietal cells to generate hydrochloric acid, decreasing the gastric pH, and subsequently stimulating peptic ulcer formation. It is estimated that 0.1% of patients with a primary duodenal ulcer have a gastrinoma. Furthermore, of patients who develop recurrent peptic ulcers after antiulcer surgery, it is estimated that 2% harbor gastrinomas.

Presentation

The symptoms of a gastrinoma are directly attributable to the release of gastrin. The majority of patients have peptic ulcer disease and its attendant symptoms of abdominal pain. Additionally, patients with a gastrinoma will often have complaints of diarrhea. In fact, some degree of diarrhea is present in approximately one-half of patients. Up to 10% of patients may have diarrhea as their only presenting symptom. Symptoms of gastroesophageal reflux may also be elicited in about one-half of patients.

Gastrinomas are sporadic in approximately 75% of cases. In the other 25% of cases, however, gastrinoma occurs in association with multiple endocrine neoplasia syndrome (MEN). Specifically, it is associated with MEN I syndrome which features pituitary neoplasia and parathyroid hyperplasia in addition to a pancreatic endocrine neoplasm. Approximately 50% of gastrinomas are malignant.

Diagnosis

Most patients with a gastrinoma have fasting serum gastrin levels >200 pg/mL. Serum gastrin levels >1000 pg/mL are strongly suggestive of a gastrinoma. However, demonstration of elevated serum gastrin levels is not sufficient to make the diagnosis of a gastrinoma as there are other causes of hypergastrinemia.

Ulcerogenic causes of hypergastrinemia include antral G-cell hyperplasia, gastric outlet obstruction, retained gastric antrum syndrome, and gastrinoma. Nonulcerogenic causes include atrophic gastritis, pernicious anemia, previous vagotomy, renal failure, and short-gut syndrome.

After confirming hypergastrinemia, the next step diagnostically is gastric acid analysis. To obtain accurate data patients must abstain from antisecretory medications prior to this analysis. Basal acid output >15 mEq/hr in nonoperated patients, >5 mEq/hr in patients with previous vagotomy, or a ratio of basal to maximal acid output >0.6 supports the conclusion that hypergastrinemia is associated with hypersecretion of gastric acid. Once this is determined, provocative testing with secretin may be performed to differentiate among the various causes of ulcerogenic hypergastrinemia. Secretin stimulation is done in the fasting state by obtaining serum samples for gastrin in the basal period and after secretin administration in 5-minute intervals for 30 minutes. Increase in serum gastrin levels of >200 pg/mL above the basal level supports the diagnosis of gastrinoma. After biochemical diagnosis is made, localization studies are then conducted. The initial imaging technique recommended for the identification of a gastrinoma is a dynamic abdominal CT scan with 5-mm pancreatic cuts and both oral and intravenous contrast.

There are a number of clinical situations in which there is a heightened suspicion of a gastrinoma. Serum gastrin levels should be obtained in patients with peptic ulcer disease when ulcers are recurrent on antisecretory therapy or when ulcers fail to heal on appropriate medical therapy. Postoperative ulcers, post-bulbar duodenal ulcers, peptic ulcers associated with diarrhea, and family history of peptic ulcer disease should also stimulate one to consider measuring serum gastrin levels. Other indications for measuring serum gastrin include prolonged undiagnosed diarrhea, observation of prominent gastric rugal folds, and the presence of other pancreatic endocrine tumors. MEN kindred should also be screened.

Management

Pharmacological control of gastric acid hypersecretion is attempted in the period preceding surgery. Omperazole (20-200 mg a day) is considered the initial drug of choice.

Most gastrinomas are found to the right of the superior mesenteric vessels, in the head of the pancreas or duodenum in the so-called gastrinoma triangle. Intraoperative ultrasound may be employed to help with identification of the tumor. In addition, transillumination via intraoperative endoscopy can help to identify small tumors of the duodenum.

Surgical Management

Primary intrapancreatic tumors less than 2 cm and located away from the main pancreatic duct may be treated with enucleation. Otherwise, deep, large, or malignant intrapancreatic tumors may need to be resected by pancreaticoduodenectomy. Gastrinomas of the duodenal wall can be locally excised with primary closure of the duodenal defect. Up to 35% of patients who undergo exploration for gastrinoma with curative intent have been rendered eugastrinemic at follow-up. When only those patients thought to be successfully resected at the time of surgery are considered, cure rates are about 60-70%. In patients with associated MEN I syndrome, omeprazole should be used to control gastric acid secretion while surgical therapy of hyperparathyroidism is performed first. MEN I patients frequently demonstrate multiple gastrinomas and overall cure rates are lower than with sporadic forms.

If preoperative localization demonstrates the tumor in the gastrinoma triangle but no tumor can be found intraoperatively, there are several surgical strategies. Total gastrectomy is one option. The introduction of omeprazole has drastically reduced the need for this approach.

This approach leaves behind tumor with the potential for growth and metastasis. For this reason, blind pancreaticoduodenectomy to include the pylorus has been advocated by some. Some patients are rendered eugastrinemic by this approach. In addition, in a small number of cases a pathologically confirmed gastrinoma can be demonstrated in the surgical specimen.

Findings of unresectable hepatic disease are confirmed by biopsy of liver lesions. If the patient has confirmed hepatic metastases, exploration and resection are not indicated and the patient is managed with antisecretory therapy. If the patient suffers complications of disease due to lack of efficacy of antisecretory medication than total gastrectomy may be considered. Objective response to chemotherapy is < 50% and has not been shown to improve survival. Hormonal therapy with octreotide has been reported to improve symptoms, decrease hypergastrinemia, and reduce hyperchlorhydria in patients with metastatic gastrinoma.

11

CHAPTER 12

Small and Large Intestines
Steven R. Casos and William R. Wrightson

Appendicitis
Steven R. Casos

Background
Fitz described the natural history of appendicitis as early as 1889. That same year, McBurney gave his classic treatise on the anatomy of appendicitis before the New York Surgical Society, leaving the legacy of McBurney's point indelibly inscribed upon every medical student since.

Epidemiology
The lifetime risk of developing acute appendicitis is 7% and hasn't changed since it was first characterized. Mortality is low, less the 1%, except in the elderly and pediatric populations. Mean age at time of surgery is 25.5 years. There is a slight predominance of men affected vs. women, up to 67% in some studies. Negative appendectomy rates traditionally 20-30% are improving slightly, but perforation rate still remains at near 20% despite advances in technology.

Pathophysiology
Obstruction of the appendiceal lumen is the inciting factor in appendicitis, and leads to perforation and peritonitis. After the appendiceal orifice is obstructed, bacterial overgrowth of the distal lumen takes place. Intraluminal pressure rises according the Laplace's law. This results in venous hypertension, which perpetuates the cycle by contributing to wall thickening. Once the venous pressure exceeds capillary oncotic pressure, the appendix becomes ischemic, necrotic, and eventually perforates. This sequence of events occurs over 24-36 hours. Inflammation of appendiceal lymphoid tissue results in the majority of cases of appendicitis, about 60%. This can be caused by something as simple as gastroenteritis or may be a manifestation of more advanced colonic disease such as Crohn's.

Anatomy
The appendix averages from 5 to 20 cm, with an average length of 9 cm. McBurney's point is defined as the area under a single finger that lies 1.5 to 2 inches from the ASIC along a straight line from that anatomical landmark to the umbilicus.

Diagnosis
Diagnosis consists chiefly of an accurate history and physical examination, with careful attention paid to the sequence of events. Pain almost always precedes nausea

Current Concepts in General Surgery: A Resident Review, edited by William R. Wrightson.
©2006 Landes Bioscience.

and vomiting and patients that state they are hungry (hamburger sign) almost invariably are not suffering from acute appendicitis.

Radiology

In most cases of appendicitis, radiographs are not necessary. Radiologic examinations are reserved in cases where ambiguity exists, or where the morbidity of the operation (including anesthesia) would be poorly tolerated by the patient.

Plain films

A fecalith is present in < 15% of cases. Free air from perforation is seen 1% of the time.[1] Overall a very poor study.

Ultrasound

Most effective in young females of child-bearing age in the evaluation of adnexal disease which is high on the differential. Some studies suggest that U/S is no better than history and physical alone, so outside the realm described previously, it has limited utility.

Computed Tomography

Superior in both pediatric and adult populations in elucidating equivocal cases of appendicitis. It has a sensitivity ranging from 96-100%, a specificity of 89-97%, a PPV of 92-97%, and a NPV of 95-100%. CT scan of the appendix has been shown to decrease the negative appendectomy rate by 13% and reduce perforations by 8%.[4] Some centers cannot perform the 0.5 to 1 cm cuts on their scanners that are required to make this exam useful.

Radionuclide

Some early studies suggest that a sensitivity and specificity of >90% can be achieved. However the added expense of about $500 and the delay in acquisition of 5 hours rarely justify this novel radiographic approach.

Management

The goal of the surgical approach to appendicitis is simple—early diagnosis with resection of an acutely inflamed appendix prior to perforation, with a minimum of "negative" appendectomies.

Treatment

Open Appendectomy

A transverse Rocky-Davis or the classical McBurney skin incision is made in the RLQ over the area of maximal tenderness. If purulent or cloudy peritoneal fluid is encountered, it should be sent for culture and sensitivity. The appendix is identified at the confluence of the taeniea coli, and the mesoappendix is clamped and divided. A silk purse string suture is placed at the base of the appendix, which is then clamped, ligated with catgut, and divided sharply. The appendiceal stump can be cauterized either chemically or electrically (dealer's choice), and "dunked" into the cecum. The fascia is closed, and the skin also except in cases of perforated appendicitis.

If the appendix is perforated, historical management has been either delayed primary closure or primary closure with drainage. This becomes more an issue in

12

the pediatric population where follow-up and cosmesis play a larger role. Serour et al suggests that primary closure with triple antibiotic therapy for 7-10 days results in a wound infection rate of 6% for children after perforated appendicitis.[4] Fischer suggests that primary closure should include the placement of either a Penrose or closed suction drain.[3]

When a normal appendix is encountered, a limited exploration is warranted to rule out nearby pathology. In all cases except for IBD, the appendix should be removed to eliminate the possibility of confusion in future cases of RLQ pain. If an appendix is removed in the presence of active IBD, a fecal fistula may ensue.

Laparoscopic Appendectomy

One randomized study suggests that even though hospital stay was about the same, patients undergoing laparoscopic appendectomy returned to work in 7 versus 10 days. They also had fewer wound infections. However, laparoscopy was associated with a greater number of intraabdominal abscesses (5% versus 1%) and a longer operating time (60 versus 40 minutes). Finally, almost a fourth of 285 patients randomized to laparoscopy required conversion to open appendicectomy. Nonetheless, the patients who underwent laparoscopy were more pleased with their cosmetic surgery.[4] Another study suggests that laparoscopic appendectomy at least had no obvious disadvantages.[4]

In defense of laparoscopy, it has proved its worth in certain circumstances, for example in women of child-bearing age, due to its increased diagnostic value. Additionally, in obese or heavily muscles individuals where larger incisions and excessive retraction may be required, laparoscopy has turned out to be the preferred modality for many.[4]

Antibiotics

The administration of antibiotics is generally not disputed, but the length of treatment is. For perforated appendicitis, some surgeons will use extended spectrum synthetic penicillins. Others will use ampicillin, gentamycin, and metronidazole. Nevertheless, monotherapy with a second generation cephalosporin is more economical and equally efficient, even in cases of perforated appendicitis with abscess formation. A total of 3 days of antibiotic therapy above and beyond the point where the patient is no longer febrile or has a leukocytosis is sufficient.

12

Appendiceal Mass and Abscess

A mass is a palpable conglomeration of inflamed tissues, including the appendix and adjacent viscera. CT scan of the abdomen and appendix can delineate a phlegmon versus an abscess, the treatment of which are distinct.

A difference of opinion revolves around the necessity of an operative approach or a more conservative regimen. A conservative approach with antibiotics, the so-called Ochsner method, is an acceptable course of treatment. Ochsner defended his method based on the following three principles: first, it is more difficult to remove the appendix when a phlegmon is present; second, one can always revert to an operative approach if the patient deteriorates; third, conservative treatment works in > 80% of cases.[4] A caveat however, the conservative approach requires an extended hospital stay initially, not to mention the interval appendectomy that will be performed at a later date.

Table 12.1. Mortality of perforated appendicitis

Age Group	Operations	Deaths Within 30 Days	Case Fatality Rate per 1000 Operations
0-9	9,756	3	0.31
10-19	37,098	3	0.08
20-29	27,054	2	0.07
30-39	15,664	3	0.19
40-49	10,937	9	0.82
50-59	6,534	14	2.14
60-69	5,160	37	7.17
70-79	3,757	97	25.86
80-89	1,407	96	68.23
90-99	140	23	164.29
TOTAL	**117,424**	**287**	**2.44**

An abscess in the RLQ should be treated with percutaneous drainage and concomitant IV antibiotics. As it resolves, an interval appendectomy can be entertained, usually at least 3 months after the attack. It has been shown that of the patients treated nonoperatively for abscess as well as phlegmon, 5% will fail this approach, and up to 40% will return within a year with recurrent acute appendicitis requiring appendectomy.[4]

Prognosis and Outcomes

Although perforation rates have not decreased over the past 70 years, mortality has decreased from 26% to less than 1% over the same period. Most of the morbidity and mortality associated from appendicitis is suffered by the very young and the very old.

A retrospective review found a perforated appendix rate of 20%. Overall mortality was only 0.24%, but of the deaths reported, 93% occurred in the age group >50 years of age (Table 12.1). Of the patients with perforated appendicitis, the percentage of deaths in the same age group increased to 97%. The negative appendectomy rate was a consistent 22%.[4]

A large Department of Defense Study further characterized those most likely to perforate. In their study of almost 5000 appendectomies performed in DoD hospitals worldwide over a one year period ending 31 Jan 1993, the perforation rate was 21%. They were more likely to occur in the age group <8 years of age (38%) and in those >45 years of age (49%) than in those in-between (18%). Negative appendectomy rate in this study was 13%, and there were only 4 deaths (0.08%).[5]

References

1. Greenfield LJ et al. Surgery: Scientific principles and practice. 2nd ed. Philadelphia: Lippencott-Raven, 1997:1251-1261.
2. Rutkow IM. Surgery: An illustrated history. St. Louis: Mosby, 1993:489.
3. Schein M, Wise L. Controversies in surgery. Berlin: Springer-Verlag, 2001:4:143-165.

Colon and Rectal Cancer

William R. Wrightson

Background

Colon cancer is the second leading cause of cancer deaths in the United States. There is a 5% risk in the US of developing colorectal cancer. Advances in the understanding and diagnostic screening of this disease has made an impact on overall survivability of this type of cancer. As many as 10% of all colorectal cancers can be found by digital rectal exam with 3-6% of patients having a synchronous lesion in the colon. The majority of colorectal cancers remain sporadic, 10% genetic or familial.

Genetics

Hereditary nonpolyposis colorectal carcinoma (HNPCC) consist of two main types. Lynch I is autosomal dominant while Lynch II is associated with ovary, breast, stomach cancers. It is suggested that colorectal cancer develops from a progression from benign adenomas which develop genetic mutations or loss of heterozygosity (LOC). Various genes have been investigated as participants in that progression (Table 12.2).

The starting point is a mutation in the APC gene, a tumor suppressor gene. Recent research suggests a link between loss of downregulation of beta-catenin as the central mediator in this process. The DCC gene is also involved in the process. Patients with loss of DCC have a poorer prognosis compared to those patients with an intact gene.

Etiology

Many studies have investigated the relationship between diet and cancer. It has been suggested that high fat diet is associated with cancer while low fiber diets lower risk. Other factors influencing the development of cancer include advancing age (>40 years old), positive family history, ulcerative colitis for >10 years (1-2% risk/year).

Screening

Screening consisted of yearly DRE and fecal occult blood tests with flexible sigmoidoscopy at 40 years old and every 3-5 years after. Much controversy exists regarding the use of colonoscopy as a screening test rather that flexible sigmoidoscopy.

12

Table 12.2. Genes involved in colon cancer

ras gene mutation (oncogene)
Chromosome 5q – APC
Chromosome 17p – p53 (tumor suppressor gene)
Chromosome 18q – DCC (deleted in colorectal carcinoma)
Chromosome 2p – hMSH2
Chromosome 3 – hMSH1

In 1998 Medicare changed their reimbursement guidelines to allow for coverage of screening colonoscopy. A recent study found that patients with small adenomas found on flexible sigmoidoscopy had proximal neoplasms found in 29% of patients.

Virtual colonoscopy in on the horizon to provide an early screening test that is clearly less invasive and possibly more acceptable to the public. Early studies suggest a good sensitivity for 5 mm sized polyps; however, more recent work suggests this to be more on the order of 67%. Further research is warrants and as rendering technology improves so too will the images.

Diagnostic Studies

Besides the reliable standard of care digital rectal exam which can identify as many as 6% of all colorectal cancers there is a significant arsenal of diagnostic tests available. Occult blood on DRE is associated with colon cancer in 10% on patients. Colonoscopy provides the optimal way to evaluate the entire colon and obtain biopsies for tissue diagnosis.

Colon Polyps
- Tubular (5% malignant)
- Tubulovilous (25% malignant)
- Villous (50% malignant)

Size and Risk of Malignancy
- 1% of polyps <1 cm are malignant
- 50% of polyps >2 cm are malignant.

Staging Work up

Physical exam to assess for bone pain (bone scan if positive), mental status changes (CT of head). Chest x-ray is used to evaluate for evidence of lung metastasis. Most now advocate CT scans of the abdomen and pelvis to assess extent of the cancer. This can help to determine involvement of adjacent structures or evidence of distant metastasis such as liver. The presence of liver metastasis does not preclude resection and may provide an opportunity to do a single stage resection or ablation of the metastasis and primary (Table 12.3).

Management

Table 12.4

Table 12.3. Colon cancer staging

Stage	Description
A	confined to the mucosa
B1	into but not through the muscularis propria
B2	through the muscularis propria
C1	into but not through the muscularis propria with positive nodes
C2	through the muscularis propria with positive nodes
D	metastasis

Table 12.4. Surgical procedures based on cancer location

Tumor Location	Procedure
Cecum and ascending colon	Right hemicolectomy. Resect distal ileum to mid transverse colon with ileocolic, right and middle colic with mesentery
Left transverse colon to splenic flexure	Extended right hemicolectomy or transverse colectomy. Resect transverse and proximal descending colon with middle and left colic arteries
Descending and sigmoid colon	Left hemicolectomy or sigmoid colectomy. Resect from splenic flexure to rectosigmoid with left colic and sigmoidal arteries
Proximal third of rectum 5-10 cm from anal verge	Anterior resection, low anterior resection or abdominal-sacral resection
Distal third of rectum	Abdominoperineal resection (Miles procedure) with sigmoid colostomy. Low anterior resection with reanastomosis if >4 cm of rectal tissue remains. Sphincter saving operations are gaining popularity

Obstructive Lesions and One Stage Operations

Patients with evidence of colonic obstruction were usually managed with a two stage operation including colostomy and subsequent takedown. Current research suggests that on-table lavage of the colon results in anastomotic leak rates comparable to preoperatively bowel prepped cases. The end result of reduced fecal load and bacterial counts following lavage makes primary anastomosis possible.

No Touch

Several studies have evaluated the no touch technique in colon cancer surgery. This requires ligation of the lymphovascular supply prior to mobilization of the colon and mass. These early studies showed a survival of 58% compared to 28% for no touch and conventional surgery respectively. More recent studies of the presence of tumor cells in the portal vein during and following surgery found that 73% of patients in the conventional group had evidence of tumor showering verses 14% in the no touch groups.

Adjuvant Therapy

New strategies addressing the postoperative and adjuvant therapy for colorectal cancer are evolving. The latest research has produced several new protocols. CPT-11 recently gained approval for colorectal cancer. There was an overall 18% response rate. In addition improvements to traditional therapy have been introduced with oxaliplatin added to 5-fluorouracil. Response rates of 46-69% have been demonstrated.

Postoperative radiation and chemotherapy for with transmural and/or node positive recal cancer has become standard. This has decreased local recurrence rates and improved overall disease free survival. A study from Memorial Sloan Kettering Cancer Center report preoperative radiation with 5-FU and leukovorin for patients with resectable transmural lesions showed 22% complete clinical response. Sphincter preserving surgery was also possible in 85% of patients thought to require APR.

The NIH has recommended adjuvant therapy for Stage III colon and rectal cancers. This has shown a definite survival benefit as well as a reduction in recurrence. Use of adjuvant therapy for Stage II remains controversial.

Current Research

New evidence suggests a role for anti-inflammatory drugs in the treatment and prevention of colon and rectal cancers. Patients with familial adenomatous polyposis who took sulindac had significant reductions of the number and size of their polyps compared to controls. New directed agents against COX-2 have shown more effect and may be important in prevention. Case controlled studies show a definite reduction in the risk of colon cancer.

Gene therapy with delivery of p53 to tumor cells has been seen in phase 1 and II trials. To this point the results have been less than promising. As delivery techniques improve so too will this form of therapy.

Follow-Up

Typical follow-up following colorectal cancer surgery is done with the goal of identifying early recurrence. Most cancers will reoccur with in 18 months and therefore the closest surveillance in during this time period. Current recommendations are: CEA every 2 months for 2 years then every 4 months for 2 years then annually, colonoscopy within the first 2-3 months then annually, LFTs every 3 months for 2 years then every 6 months for 2 years then annually and CXR every 6 months for 3 years then annually.

References

1.   Chung-Faye GA, Kerr DJ. Innovative treatment for colon cancer. Br Med J 2000; 321:1397-1399.
2.   Wu JS, Fazio VW. Colon cancer. Diseases of the colon and rectum 2000; 43:1473-1486.

12

Fistulas and Their Management

William R. Wrightson

Background

The most frequent source of development of a bowel fistula is previous surgery. As a postoperative complication, enterocutaneous fistulas can be disastrous. Advances through the mid 1900s resulted in a decrease in mortality rates from 45% to 15%. Current management strategies including resuscitation, liberal use of antibiotics, nutritional support and wound care have improved patient outcome.

Etiology

Common sources for fistula formation include laparotomy for infected pancreatic necrosis, left colonic resections, intestinal resection for gangrene, closure of colonic perforation, adhesiolysis, hysterectomy and ovariectomy. Colonic injuries were also discerned following laparoscopic pelvic surgery. Cautery burns to colon occur despite attention to detail with grazing burns progressing to an area of ischemia and perforation. Recently, colonic fistula have been reported following percutaneous

Table 12.5. Enterocutaneous fistulas

Cause of Enterocutaneous Fistula	Rate
Postoperative complication	82%
Inflammatory bowel disease	6.3%
Cancer	3.8%
Infectous	2.5%
Iatrogenic	2.5%

drainage of abdominal abscesses. Spontaneous fistula formation due to disease process occurs usually into an internal organ. Thus colovesical fistula or colovaginal fistula may complicate colonic diverticulitis, carcinoma or Crohn's disease (Table 12.5).

Factors involved in the development of fistulas include are excess tension or ischemia at bowel anastomosis. Abscesses, distal bowel obstruction and acute necrotizing pancreatitis may also contribute to fistulization. Patient comorbidities predispose to fistula are catabolic illness, severe malnutrition (with serum-albumin <2 g/dL), marked diabetes mellitus, morbid obesity, prolonged steroid therapy.

Fistula Classification

Fistulas are classified by daily outputs and location (Table 12.6). The greater the output the more severe the fistula and the less likely it will close spontaneously. The greater the distance of the fistula tract is directly proportional to the likelihood of closure. That is a longer tract (>2 cm) is more likely to close spontaneously.

Complications of Fistulas

Colocutaneous fistulas create significant morbidity.

- Fluid and electrolyte imbalance due to loss of intestinal contents
- Bacterial contamination of usually sterile areas can result in peritonitis and sepsis
- Bypass of a variable length of the intestine leads to a loss of functional intestine and impaired digestion and absorption
- Wound infection
- Skin excoriation by digestive enzymes

The extent of these problems is governed by level, location and magnitude of the fistula. In addition to these, there will always be problems due to underlying disease and comorbidity. The problems are maximum with small intestinal fistula a minimum with left colonic fistula.

12

Table 12.6. Fistula output in 24 hours

Classification	Output
High output	>1000 mL per day
Medium output	500-1000 mL per day
Low output	<500 mL per day

Diagnosis

One of the most essential studies to be carried out is a fistulogram to assess the anatomy of the fistula. If a fistulogram in not possible other means of assessing the fistula can be carried out. The possible localization studies are:

- Gastrografin per rectum - usually required
- Oral gastrografin studies (infrequently required - done if rectal gastrografin study is inconclusive)
- Gastrografin study through the fistula.

The aims of these studies are to locate the level of the fistula, define its type if possible, assess the completeness or otherwise of the rupture, and detect a distal obstruction if present. X-ray films at 3, 6, and 12 hours after transluminal contrast studies may show a delayed filling of an abscess cavity.

For colovesical fistula, cystoscopy is mandatory. Colonoscopy is essential in spontaneous fistula to assess the colonic pathology but is best avoided in the postoperative fistulae.

Management

Management considerations are:

- Resuscitation and fluid and electrolytes imbalance
- Nutritional support
- Medical treatment (antibiotics, octreotide, H_2 blockers)
- Skin protection
- Surgery and its timing

Initially adequate fluid and electrolyte administration is necessary to correct dehydration, achieve hemodynamic stability and maintain renal function. Skin is protected by the dual use of stoma bag and stoma adhesive / Karaya paste / zinc peroxide paste. The stoma bag allows measurement of fluid losses and, to an extent, the electrolytes and hence better maintenance of fluid and electrolyte balance. It also provides a measure of the response to therapy.

Fistulae through drainage tube sites and which are away from (a) main wound, (b) scars of previous surgery, (c) umbilicus, (d) bony prominences and accept the stoma bag snugly. In situations where a stoma appliance will not function properly, ingenuity of clinician and of the stoma nurse as well as versatility of appliance are necessary. Placement of DuoDerm around the fistula with red rubber catheters irrigating and draining the area are also useful.

Medical Management

Octreotide (synthetic analogue of somatostatin) in doses of 100 mg (SC inj.) has been shown to decrease pancreaticobiliary and gastrointestinal secretions. Although its effectiveness in fistula closure remains unproven. It may even lead to heading of the fistula. Much controversy still exists concerning the use of octreotide especially in light of the excessive cost and uncertain benefits.

H2 blockers and other antacids provide substantial decrease in fistulous output by reducing the stimulatory effect of acid gastric secretion on the bowel.

Antibiotics are frequently utilized because more than 75% of the patients have a local infection with or without abscess. Guidelines are obtained by culture and sensitivity of swab from wound if this is infected. Patients with proven invasive fungus infection are also administered fluconazole or other appropriate antifungals.

Percutaneous Drainage

Ultrasound guided drainage or CT scan should be carried out to rule out local collections or abscesses. If found these should be aspirated or therapeutically drained by a pigtail catheter through a window devoid of bowel and vital structures. A complete microbiological examination of the fluid is essential. Technical inability to drain such a collection or an abscess is an indication for early surgery.

Nutritional support is the single most important therapy which in recent times has contributed to a decrease in morbidity/mortality and even led to spontaneous closure of many low output and some moderate output fistulae. It is started only after achieving hemodynamic stability and normalization of electrolytes and renal status. Higher the fistula in the intestinal tract lesser is the tolerance and effectiveness of enteral nutrition. Even if the nutritional support does not heal the fistula, it does decrease the fistulous effluent, minimizes local infection, and increases tissue integrity. All these ensure a better outcome of the reparative operation. Patients with a serum albumin of less than 2.5 g/dL have a mortality of 64% and spontaneous closure of only 23%.

Surgery

Those who require immediate surgery (in less than 24 hours) have
- proven or suspected gangrene
- severe peritonitis,
- life threatening infection or,
- total disruption of intestinal continuity

In these, apart from intestinal resection if required, the two intestinal ends are best brought out as diversion proximal stoma, and mucous distal stoma, (to be joined at a later date, as a second stage). Primary anastomosis in these circumstances invariably breaks down.

Early surgery (with 3-5 days) is required in those with
- distal obstruction
- specific disease
- colovesical fistula and
- an abscess or collection which cannot be adequately drained percutaneously under US/CT guidance

Formerly loop transverse colostomy was the preferred diversion whenever it was feasible, but during the last 5-10 years (loop ileostomy which is nearly always possible, is the diversion of choice. Ileostomies offer an advantage because the stoma bag fits better, the odor is less, interferes less with subsequent colonic resection and anastomosis, and closure is easier with less morbidity.

Outcome

The majority of low level fistula heal with medical treatment outlined earlier. Almost 15-20% of the fistula with medium output (500-1000 mL/day) will also heal, whereas majority of high output fistula will require surgery. Over all prognoses depend on several factors (Table 12.7).

Table 12.7. Factors involved in failure of fistula closure

Factor	Details
Site of fistula	Small intestinal fistula have worse prognosis (mortality over 15%) whereas the left colonic fistula has the best prognosis with practically no mortality.
Output	High output fistula has a worse prognosis and is less likely to close
Number	Those with more than one fistula (Crohn's) are less likely to close spontaneously.
Severity of presentation	Patients presenting with poor nutritional status, sepsis, pancreatic necrosis, tissue loss have a more protracted outcome.

References

1. Cameron JL. Current surgical therapy. 7th ed. Mosby Inc., 2001.
2. Sitges-Serra A, Jaurrieta E, Sitges-Creus A. Management of postoperative enterocutaneous fistulas: The roles of parenteral nutrition and surgery. Br J Surg 1982; 69:147-50.
3. Nubiola-Calonge P, Badia JM, Sancho J et al. Blind evaluation of the effect of octreotide (SMS) 20+995) a somatostatin analogue, on small bowel fistula output. Lancet 1987; ii:672-9.
4. Reber H, Roberts C, Way L et al. Management of external gastrointestinal fistulas. Ann Surg 1978; 188:460-7.

12

CHAPTER 13

Liver

Dan N. Tran and William R. Wrightson

Primary Liver Anatomy and Physiology

Dan N. Tran

Anatomy

Functionally, the liver is divided into the left and right lobes (or hemilivers) by Cantlie's line. This line is parallel and in-line with the inferior vena cava and gallbladder fossa. It is further divided into sections based on the distributions of the branches of the hepatic artery, portal vein, and bile ducts. In this way, the right lobe is divided into the anterior and posterior sections while the left lobe is divided into the medial and lateral sections (Fig. 13.1).

Portal Vein

The portal vein is formed from the confluence of the inferior mesenteric vein and the splenic vein. It is valveless and provides roughly 75% of the liver's blood supply. It is usually located behind the common bile duct and hepatic artery in the hepatoduodenal ligament. The portal vein receives almost all the blood flow from the digestive tract between proximal stomach and upper rectum, as well as the spleen, pancreas, and gallbladder

Hepatic Artery

The hepatic artery supplies about 25% of the blood to the liver. It originates from the celiac axis as the common hepatic artery. It ascends in front of the portal vein, usually to the left and behind the bile duct, and gives off the left and right hepatic arteries. The gallbladder is supplied by one or two cystic arteries that arise from the right hepatic artery.

Hepatic Veins

The main hepatic veins are the right, middle and left. The right drains most of the right lobe while the left drains the the medial segment of the left lobe and the anterior segment of the right lobe. The left hepatic vein drains the left lateral segment.

Metabolism and Primary Hepatic Function

The liver is involved in the metabolisms of:

1. carbohydrate
2. protein
3. lipid
4. vitamins
5. bilirubin
6. toxins

Current Concepts in General Surgery: A Resident Review, edited by William R. Wrightson. ©2006 Landes Bioscience.

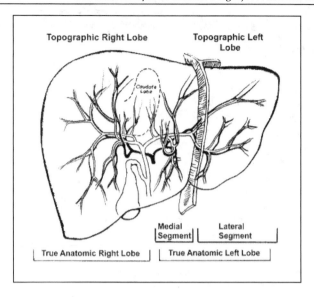

Figure 13.1. Liver anatomy.

Carbohydrate Metabolism

Dietary carbohydrates, which consists of glucose, galactose and fructose, are converted to glycogen for storage in the liver. The liver has the ability to store about 65 g of glycogen per kg of its mass. The excess carbohydrate is converted to fat. During the fasting state, when glycogen is exhausted (usually after 2 days), the liver can convert amino acids, mainly alanine, to glucose.

Bilirubin Metabolism

Bilirubin, which comes from red blood cells, is removed from the plasma by the liver. It is then conjugated with glucuronide and secreted in bile. Once secreted into the intestine, bacteria reduces the bilirubin to mesobilirubin and stecobilinogen, which are excreted in stool and urine.

Protein Metabolism

The liver is the only organ that produces serum albumin as well as synthesizing most of the urea in the body. It also has the ability to convert amino acids to glucose, and produces fibrinogen, prothrombin, and other blood clotting factors.

Lipid Metabolism

Excess dietary carbohydrate is converted to fat by the liver. The export of triglycerides is dependent of very low density lipoproteins (VLDL) which are synthesized by the liver. In the liver, 90% of cholesterol is synthesized de novo from acetyl-coenzyme A, a substrate for fat synthesis.

Vitamin Metabolism

The liver is involved in the activation of vitamin D, converting it to 25 hydroxycholecalciferol which undergoes 1- hydroxylation in the kidney to become

Table 13.1. Causes of cirrhosis

Hepatitis C & B	Autoimmunity
Alcohol	Wilson's disease
Hemochromatosis	Alpha 1-antitrypsin deficiency
Cholestasis	

the active form. It is also an uptake as well as storage site for the fat soluble vitamins, A,D,E, and K. It is the only storage place for vitamin A.

Metabolism of Drugs and Toxins

The cytochrome P-450 also known as the phase I biotransformation of drug and toxins takes place in the liver.

References

1. Sabiston Jr DC. Texbook of surgery. 15th ed. Philadelphia: WB Saunders, 1997.
2. Schwartz SI, Shires GT, Spencer FC. Principles of surgery. 7th ed. New York: McGraw Hill, 1999.

Cirrhosis and Portal Hypertension

Dan N. Tran

Background

Cirrhosis of the liver is responsible for 85% of portal hypertension in the US. First described by Laennec in 1826, cirrhosis has an incidence of 3-5% and is a significant cause of mortality and morbidity.

Etiology

Cirrhosis results from fibrosis and nodular regeneration after hepatic necrosis. The two most common causes of cirrhosis in the United States are alcoholic cirrhosis and posthepatic cirrhosis. The end result of cirrhosis is liver failure and portal hypertension. Portal hypertension leads to hepatic vascular resistance causing variceal hemorrhage while liver failure leads to encephalopathy and ascites (Table 13.1).

Cirrhosis is categorized as:
1. prehepatic
2. intrahepatic
3. post-hepatic

Intrahepatic portal hypertension may be further classified as presinusoidal, sinusoidal, or postsinusoidal. Portal vein thrombosis is the most common cause of prehepatic portal hypertension. Schistosomiasis is the most common cause of presinusoidal intrahepatic portal hypertension, while alcoholic cirrhosis is the most common cause of sinusoidal hypertension. Budd-Chiari syndrome is one cause of postsinusoidal hypertension.

Patients with cirrhosis often have elevated liver enzymes. These can include aspartate aminotransferase (AST) and alanine aminotransferase (ALT) as well as alkaline phosphatase, and gamma-glutamyl transpeptidase (GGT).

Elevation of the hepatocellular enzymes to more than three times their normal level is indicative of significant, ongoing hepatocellular necrosis, which is often present

13

Table 13.2. Hepatitis panel

	HBsAg (surface Ag)	Anti-HBsAg	HBeAg (core related Ag)	Anti-HBeAg
Acute HBV	+	-	+	-
Chronic HBV Infective	+	-	+	-
Chronic HBV Low infectivity	+	-	-	+
Recovery stage	-	+	-	+
Immunized	-	+	-	-

in patients with alcoholic hepatitis and chronic active hepatitis resulting from a variety of causes (Table 13.2).

Cirrhosis is most accurately diagnosed by percutaneous liver biopsy, and the Child-Pugh classification has been widely used to measure hepatic reserve in patients with cirrhosis (Tables 13.3, 13.4).

Complications of Cirrhosis and Portal Hypertension

Variceal Hemorrhage

Thirty percent of all deaths from cirrhosis are due to esophageal and gastric varices. While about 20% of portal hypertensive bleedings are from gastric varices, up to 80% are from esophageal varices. Bleeding from esophageal varices occurs when the portal pressure is greater than 10-15 mm Hg.

Management

The treatment for acute bleeding from esophageal varices includes initial resuscitation. When the patient is stable, endoscopic sclerosis or ligation are effective means of controlling the bleeding. Drugs such as vasopressin and octreotide are also effective as temporizing measures in controlling the bleeding. Sengstaken-Blakemore balloon tamponade can be used to stop exsanguinating hemorrhage when the patient is nonresponsive to endoscopic or pharmacologic attempts. TIPS (transjugular intrahepatic portosystemic shunt) can also be used should initial treatments fail.

Emergency surgery is indicated when the above mentioned treatments have failed to stop the bleeding. The most effective emergent procedure is the portacaval shunt as it can rapidly decompresses the portal venous circulation.

Table 13.3. Child-Pugh classification

Criteria	A	B	C
Bilirubin	< 2	2-3	> 3
Albumin	> 3.5	2.8-3.5	< 2.8
Prothrombin time	1-3	4-6	> 6
Ascites	None	Mild	Moderate
Neurologic disorder	None	Mild	Severe

Table 13.4. Operative mortality rates in association with the Child-Pugh classification

A	5%
B	15%
C	> 25%

Prevention of Recurrent Hemorrhage

Prevention of recurrent hemorrhage from esophageal varices include:
1. pharmacotherapy with beta blockers
2. repeated sclerotherapy
3. transjugular intrahepatic portosystemic shunt
4. portosystemic shunts
5. hepatic transplantation

Portosystemic Shunts

Portosystemic shunts can be classified as nonselective, selective, or partial (Table 13.5).

Nonselective shunts are complicated by frequent (30-40%) postoperative encephalopathy and accelerated hepatic failure. The manifestations of encephalopathy include altered level of consciousness, intellectual decline, personality changes, tremor, and asterixis. The cause of encephalopathy may be attributed to toxins such as ammonia, mercaptans, and gamma-aminobutyric acid.

Treatment for hepatic encephalopathy may include lactulose and or oral neomycin.

Ascites

Another complication of cirrhosis is the formation of ascites. It is usually an indicator of advanced cirrhosis. Ascites occur as a result of splanchnic congestion resulting in transudation of fluid into the interstitial space. Treatment is mainly medical that consists of dietary salt restriction (20-30 mEq/day) as well as diuretic therapy (spironolactone).

13

References
1. Sabiston Jr DC: Texbook of Surgery. 15th ed. Philadelphia: W.B. Saunders, 1997.
2. Schwartz SI, Shires GT, Spencer FC: Principles of Surgery. 7th ed. New York: McGraw Hill, 1999.

Table 13.5. Portosystemic shunts

End-to-side (Eck fistula)	Nonselective
Side-to-side portalcaval	Nonselective
Large-diameter interposition graft	Nonselective
Splenorenal	Nonselective
Distal spenorenal	Selective
Small-diameter interposition graft	Partial

Table 13.6. Causes of pyogenic liver abscess

Cholecystitis	Regional enteritis
Choledocholithiasis	Pelvic inflammatory disease
Biliary stricture	Pancreatic cancer
Diverticulitis	Biliary cancer
Trauma	Colon cancer
Appendicitis	Portal pylephlebitis
CMV and candida infection after liver transplantation	

Liver Abscesses

Dan N. Tran

Background

Medical management is the cornerstone of therapy in amebic liver abscess while early intervention in the form of surgical therapy or catheter drainage and parenteral antibiotics is the rule in pyogenic liver abscess.

Pyogenic

Incidence

Eighty percent of liver abscesses in the United States are pyogenic while 10% of hepatic abscess are from amebiasis. Patients with pyogenic hepatic abscesses are usually in the 40-60 year-old age group. Patients with amebic abscesses are usually younger, in their 20-30s.

Etiology

Appendicits was the leading cause of pyogenic abscess in the past. Biliary disease has become the leading cause today (Table 13.6).

Route of bacterial spread to the liver include:
- the portal system (most common)
- biliary tree
- the hepatic artery (bacteremia)
- direct extension from subhepatic or subdiaphragmatic infection
- directly from during trauma

Organisms most commonly involved in pyogenic liver abscess are *Escherichia coli*, Klebsiella, and Enterococcus. Bacteroides and Streptococcal species are also found.

Presentation

Patients usually present with fever, right upper quadrant abdominal pain, and malaise. Abdominal CT scan as well as ultrasound are sensitive diagnostic tools and can detect > 90% of the cases.

Pyogenic abscess is usually treated with intravenous antibiotics and either percutaneous or open drainage. In some instances, resection may be necessary. Currently, percutaneous drainage with IV antibiotics is the preferred treatment.

13

Amebic Liver Abscess

Amebic liver abscess is caused by *Entamoeba histolytica* and usually involves the right lobe.

Incidence

It is estimated that between 5 and 10% of the population have intestinal amebiasis. Amebic abscess affects males more than females in as much as a 9 to 10:1 ratio. In general, the patients are younger, with the highest incidence in the 20- to 50-year-old age group.

Etiology

Common modes of transmission of this organism are usually by contaminated drinking water or food as well as individual contact. The organism can exist in two forms, trophozoites and cysts. It is the trophozoites that are the invasive form.

Pathology

Entamoeba histolytica reaches the liver via the portal vein where it causes liver necrosis. Secondary bacterial infection can sometimes set in. When the abscess is large, it can rupture. Right lobe abscess tend to rupture intraperitoneally, while left lobe lesions can rupture into the pleural space or even the pericardium.

Presentation

Patients with amebic abscess sometimes present with a history of recent diarrhea.

Diagnosis

Ultrasound and CT are also useful tools in visualizing amebic abscesses. Distinctions between pyogenic and amebic abscess, however, cannot be made with either of these imaging modalities. Serum antibody against *Entamoeba histolytica* are highly specific and this test is useful for distinguishing between the two types of abscesses.

Management

Amebic abscess is effectively treated with metronidazole, chloroquine, or dehydroemetine.

References

1. Sabiston Jr DC. Texbook of surgery. 15th ed. Philadelphia: WB Saunders, 1997:3.
2. Schwartz SI, Shires GT, Spencer FC. Principles of surgery. 7th ed. New York: McGraw Hill, 1999.

13

Benign and Malignant Hepatic Neoplasms

Dan N. Tran

Hemangiomas

Most hemangiomas are small and do not cause symptoms. However, they can rarely become large and undergo spontaneous rupture. The decision to resect hepatic hemangioma depends of the size of the tumor and the symptoms that it produces (Table 13.7).

Table 13.7. Distinctions between adenomas and focal nodular hyperplasia

	Adenoma	FNH
Occur primarily in young women	Yes	Yes
Associated with oral contraceptive use	Yes	Yes
Prone to hemorrhage and necrosis	Yes	Rare
Malignant change is possible	Yes	Rare
Composed of hepatocytes	Yes	yes

Cysts

These are usually benign and can be solitary or multiple. Symptoms of large cysts are increased abdominal girth, pain, and rarely, obstructive jaundice.

Treatment is usually by excision. However, if the cyst communicates with the biliary system a Roux-en-Y cystojejunostomy is the treatment of choice.

Malignant Cancer of the Liver

Hepatocellular Carcinoma (HCC)

HCC is the most common primary cancer of the liver in adult and is the most common malignant neoplasm worldwide.

Etiology

See Table 13.8 Etiology of Hepatocellular Carcinoma.

Presentation

Symptoms of HCC include weight loss, abdominal pain, malaise, jaundice, hepatomegaly, dyspnea, and anorexia. Over 90% of patients have metastatic disease at initial presentation and possibility for resection is limited.

Most patients with HCC secondary to viral hepatitis and cirrhosis have AFP levels above 300 to 400 ng/mL. Patients with HCC larger than 5 cm are reported to have AFP greater than 100 ng/liter.

Management

Resection can provide cure and at present is the only therapy that substantially prolongs survival, especially if the patient does not have cirrhosis. Negative prognostic factors at presentation include: low albumin, advanced age, portal vein obstruction, and AFP level (Tables 13.9 and 13.10).

Table 13.8. Etiology of hepatocellular carcinoma

Hepatitis C	Alcoholic cirrhosis
Hepatitis B	Aflatoxin
Hepatic adenoma	Other types of cirrhosis
Wilson's disease	Tyrosinemia
Glycogen storage disease	Tamoxifen
hemochromatosis	

Table 13.9. Variants of hepatocellular carcinoma

Fibrolamellar	Usually found in patients younger than 40 years old. It has a more favorable prognosis than standard HCC.
Carcinosarcoma	Has features of both HCC and spindle cell carcinoma.
Clear cell carcinoma	Sometimes difficult to distinquish from metastatic renal cell carcinoma.
Giant cell carcinoma	Composed of multinucleated and pleomorphic large cells

Metastatic Tumors

This is the largest group of malignant tumors in the liver. Metastatic disease to the liver is secondary to its rich vascular supply and extensive drainage. Twenty-five percent of patients with metastatic disease to the liver are considered resectable. Of those, only 30% will have long term survival (Table 13.11).

Diagnosis

A serum CEA concentration greater than 9 ng per mL combined with positive results of a liver imaging test predicts metastases with 98% accuracy. An elevated serum AFP concentration is reported to have a 75%.

Treatment for Colorectal Metastasis

Treatment includes:
- resection
- ablation (radiofrequency or cryoablation)
- hepatic artery infusion pump
- chemotherapy (5-fluorouracil)

Table 13.10. Other types of primary liver neoplasms

Epithelial Tumors

Hepatoblastoma	The most common malignant liver tumors in childhood. Drived from embryonic or fetal hepatocytes.
Bile duct cancer (Cholangiocarcinoma)	is associated with chronic cholestasis, cirrhosis, hemochromatosis, and congenital cystic disease of the liver. *Clonorchis sinensis* infestation is associated with more than 85% of cholangiocarcinomas in Asia.
Hepatic cystadenocarcinoma	arise from benign cystadenomas
Squamous cell carcinoma	

Table 13.11. Source of metastatic liver disease

1. bronchogenic carcinoma	5. pancreas
2. prostate	6. stomach
3. colon	7. kidney
4. breast	8. cervix

13

Table 13.12. *Major liver resections*

Right lobectomy	Resection right of Cantlie's line
Right trisegmentectomy	Resection right of umbilical line
Left lobectomy	Resection left of Cantlie's line
Left lateral segmentectomy	Resection left of umbilical line

Resection remains the treatment of choice for metastatic colorectal cancer to the liver. Resection of can provide as much as a 40% 5-year survival rate in some instances.

Contraindications to resection include:
- distant metastasis or extrahepatic disease
- portal or IVC involvement
- advanced cirrhosis
- total hepatic involvement

Liver Resection

Types of Resection

The four major types of hepatic resection are based on the lobar system of anatomy (Table 13.12).

Anatomic resection is preferred for the treatment of malignant disease as cancer cells tend to follow the portal venous pattern. Non-anatomic resections and enucleations are often reserved for cysts, benign liver mass, or for tumor debulking.

Ablation of Liver Tumors

Liver ablation includes cryoablation, radiofrequency ablation, ethanol, or acetic acid via injection. Chemotherapy arterial embolization using doxorubicin, cisplatin, or mitomycin has also been used.

References

1. Jackson PE. Baillieres best pract pres clin gastroenterol. Dec 1999; 13(4):545-55.
2. Sabiston Jr DC. Texbook of surgery. 15th ed. Philadelphia: W.B. Saunders, 1997.
3. Schwartz SI, Shires GT, Spencer FC. Principles of surgery. 7th ed. New York: McGraw Hill, 1999.
4. Sharma MP. Indian J Gastroenterol 2001; 20(Suppl 1):C33-6.
5. Sleisenger & fordtran's gastrointestinal and liver disease. 6th ed. Philadelphia: W.B. Saunders, 1998.

Liver Tumor Ablation

William R. Wrightson

Liver cancer remains a significant cause of cancer mortality. The majority of these lesions are metastatic with colorectal metastasis accounting for over 60%. Median survival of these lesions is limited with a 5-12 months reported in the literature. Ultimately, the 5 year survival with surgery is 30-45%. Unfortunately only 25-30% of patients with these lesions are resectable and 5-15% with primary hepatocellular carcinoma are resectable. Innovative strategies have been developed to improve treatment.

Cryosurgery

Cryosurgery introduced for liver tumors was riddled with postoperative complications. This process allowed small- to moderate-sized metastasis to be destroyed by cyclic freezing and thawing. The end result was a lysis of the tumor. It is believed that much of the postoperative morbidity was due to the release of intracellular contents that induced a systemic inflammatory response. In addition, cryosurgery was time consuming, with expensive bulky equipment.

Radiofrequency Ablation (RFA)

Ablation of tissue with thermal energy has been available for over 70 years but its application to the destruction of tumors is relatively recent. Localized application of thermal energy is used to heat tissue temperatures exceeding 50°C. This results in intracellular denaturation of protein and destruction of lipid bilayers. Only the tissue with RF current applied is heated to the cytotoxic temperature with surrounding tissue spared.

Technique

Intraoperative ultrasound is used to facilitate location of the tumors and assist in guiding the RFA probe into the tumor core. A RFA probe is introduced into the lesion and an array of electrodes is deployed into the tumor. After deployment, power of 50 watts is applied and increased in 10 watt increments every 60 seconds to a maximum of 90 watts. This is continued until "roll off" occurs. This indicates a drop in power output as tissue impedance increases secondary to coagulation necrosis. After a 30 second pause, the power is reapplied at 75% of the total for "roll off" and advanced as before. The probe is reapplied to encompass the entire tumor. Each cycle should ablate a sphere of 3-5 cm in diameter depending on the size of the probe used.

Application

There are multiple applications of RFA and it can be used in combination with resection and hepatic artery infusion pump.

- Multiple bilateral metastasis
- Lesions in approximation of hepatic veins, portal vein, inferior vena cava
- Lesions at lines of resection

For example, patient with extensive bilateral tumors would be considered unresectable; however a left lobectomy with ablation of the right-sided tumors is possible. The probe can be used on tumors that abut the hepatic and portal veins without damage to these structures. RFA will however ablate bile ducts and should be avoided if major compromise of the main biliary system is possible.

A study of 123 patients using RFA showed tumor recurrence in 2% of patients; however metastatic disease occurred in 28% at 15 month median follow-up. Bowles et al report a 9% local failure rate. Complication rates range from 2.4% to 10% with an average length of stay is 3 days.

References

1. Curley SA, Izzo F, Delrino P et al. Radiofrequency ablation of unresectable primary and metastatic hepatic malignancies. Ann Surg 1999; 230:1-8.
2. Bowles BJ, Machi J, Limm WM et al. Saftey and efficacy of radiofrequency thermal ablation in advanced liver tumors. Arch Surg 2001; 136:864-869.

CHAPTER 14

Biliary System

Alina D. Sholar and William R. Wrightson

Cholelithiasis: Presentation, Complications and Management
Alina D. Sholar

Background
The Roman anatomist Galen is credited with the first accurate description of the gallbladder. The first successful cholecystectomy was performed in 1882 by Langenbuch. Courvoisier, who is well-known for his description of enlargement of the gallbladder due to an obstruction of the common bile duct by pancreatic cancer, also was the first to perform a choledochotomy with stone extraction in 1890.

Epidemiology
Gallstones are a common disorder with its prevalence greatest in western societies. The Pima Indians of the Southwestern U.S. have a gallstone prevalence of about 70%. Women of reproductive age have a greater risk than men of the same age. Obesity is also a significant risk factor, as is rapid weight loss in the obese.

It is estimated that up to 80% of patients with gallstones are asymptomatic. According to epidemiologic studies, asymptomatic patients have an indolent course with 1-4% developing mild symptoms yearly, only a 2-3% rate of occurrence of serious symptoms or complications over a 20 year period, and a 20% chance of developing symptoms over their lifetime (Fig. 14.1).

Pathophysiology
Gallstones form as a result of bile salt precipitation. They are generally classified according to their composition: 70% are cholesterol stones (cholesterol + bilirubin + calcium carbonate). Others are pigment stones—a result of hemolytic disorders, parasitic infection, ileal resection, cirrhosis, biliary infection with *E. coli* and *Klebsiella*.

The clinical presentation of cholelithiasis is varied. Initially, gallstones are clinically silent and may remain so for the lifetime of the patient. However some patients will manifest symptoms at some point. The clinical presentation of cholelithiasis includes the following:
- asymptomatic cholelithiasis
- biliary colic
- chronic cholecystitis
- acute cholecystitis

Current Concepts in General Surgery: A Resident Review, edited by William R. Wrightson.
©2006 Landes Bioscience.

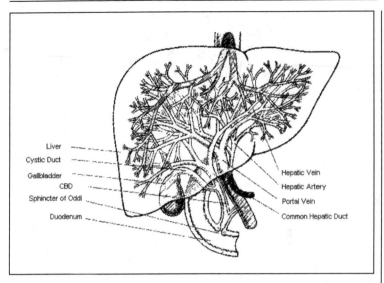

Figure 14.1. Biliary anatomy.

Complications of Cholecystitis
- hydrops
- emphysematous gallbladder
- empyema
- perforation
- obstructive jaundice
- pancreatitis
- gallstone ileus
- choledocholithiasis

Biliary Colic (Symptomatic Cholelithiasis)
Biliary colic is the pain typical of intermittent obstruction of the cystic duct by a gallstone. Patients complain of recurrent bouts of right upper quadrant, epigastric, or back pain which worsens postprandially. Nausea and emesis often accompany the pain. These episodes usually last no longer than a few hours, with relief coinciding with dislodgement of the stone. Physical exam during this period reveals mild right upper quadrant tenderness only.

Chronic Cholecystitis
Patients with chronic cholecystitis usually complain of mild, intermittent right upper quadrant pain or epigastric pain. This may be clinically indistinguishable from biliary colic, but in contrast, pathologic changes vary from minimal inflammation to widespread transmural fibrosis with contracture of the organ.

14

Acute Calculous Cholecystitis

Acute cholecystitis is the inflammation of the gallbladder following persistent obstruction of the neck of the gallbladder or cystic duct by a gallstone. The gallbladder wall becomes edematous and thickened (>4 mm) due to mucosal ischemia, impaired venous return, and subsequent release of platelet activating factor as well as leukotrienes C4 and prostaglandins. Part of the disease course is a secondary bacterial invasion of the gallbladder wall with subsequent fever and leukocytosis. Patients present with right upper quadrant pain, along with nausea and vomiting, which gradually worsens. Laboratory tests usually show only mild elevation of bilirubin levels and alkaline phosphatase, indicating biliary tree obstruction. The obstruction is usually transient; as intracholecystic pressure rises the stone is dislodged. Symptoms are then relieved and inflammation resolves. In approximately 10% of patients, the disease course is characterized by the development of complications such as empyema, perforation with resultant peritonitis, and fistulization leading to gallstone ileus.

Diagnosis and Radiologic Findings

The diagnostic study of choice is ultrasound. Ultrasonographic findings are cholelithiasis, a thickened gallbladder wall, a distended gallbladder, and pericholecystic fluid. Ultrasound diagnoses the disease 98% of the time, but if ultrasonography is equivocal, a radionuclide HIDA scan is used to determine if the cystic dust is obstructed. HIDA with nonvisualized gallbladder or decreased ejection fraction (<30%) is highly associated with cholecystitis.

Management

The patient with acute cholecystitis is initially managed with intravenous fluid rehydration and nasogastric tube decompression if vomiting is present. Antibiotics directed against gram-negative organisms are recommended to be initiated preoperatively.

After resuscitation has been optimized, cholecystectomy is indicated with the laparoscopic approach being the most favored. The timing of operation has been well-studied with trials showing that early cholecystectomy is preferable over a delayed approach. If symptoms began within 72 hours of presentation, a laparoscopic cholecystectomy should be performed at the first available elective OR time. If presentation is >72 hrs after symptom onset and the patient is responding to medical management (nasogastric suction, fluid resuscitation, antibiotics), then cholecystectomy can be delayed 4-6 weeks to allow for recovery from the acute attack. This does carry an approximately 20-50% risk of recurrence, and the patient may default after discharge. For these reasons most advocate early cholecystectomy anyway. If the patient deteriorates or does not improve, this is an indication for surgery. Also, the presence of an inflammatory right upper quadrant mass, gas in gallbladder or biliary tract, and peritonitis each call for emergency operation.

If the patient's comorbid condition deems them to be high-risk and precludes surgical intervention, a cholecystotomy tube can be placed to drain the gallbladder until the patient has recovered enough to tolerate cholecystectomy. Following resolution of the acute process, cholangiography is performed and residual stones are evacuated prior to removal of the tube.

Gallstones discovered incidentally do not require treatment, and therapy should only begin if the patient becomes symptomatic. Exceptions are all children and those patients with sickle cell disease, a nonfunctioning gallbladder, and calcified gallbladder (up to 50% associated with cancer of the gallbladder.)

Prophylactic cholecystectomy is not recommended in diabetic patients because the operative risk outweighs the benefit of cholecystectomy. In patients undergoing laparotomy for other reasons, if calculi are noted, cholecystectomy is appropriate if the primary operation is without apparent complication and exposure is adequate.

References
1. Greenfield LJ, Mulholland M, Oldham KT et al. Surgery Scientific Principles and Practice, 2nd ed. Philadelphia: Lippincott-Raven, 1997.
2. Sabiston DC, Lyerly HK. Textbook of Surgery. 15th ed. Philadelphia: WB Saunders Co., 1997.

Acalculous Cholecystitis
Alina D. Sholar

Background
Acalculous cholecystitis is classically encountered in critically ill patients. In 5% of cholecystitis cases, calculi are absent.

Epidemiology
Risk factors are prolonged ileus, long-term opiate administration, multiple blood transfusions, total parenteral nutrition, and possibly some antibiotics as well. This condition develops as a consequence of prolonged gallbladder distention, bile stasis, and sludge formation, followed by mucosal damage and vessel thrombosis. The ICU patient may present with fever or unexplained sepsis. Abdominal signs may be absent or cannot be appreciated because of the patient's clinical situation.

Diagnosis and Radiology Findings
Ultrasonography findings suggestive of acalculous cholecytitis include gallbladder distention, the presence of sludge within the gallbladder, wall thickening, and pericholecystic fluid. If the diagnosis remains unclear then HIDA scintigraphy should be performed.

14

Management
Management options are cholecystectomy or cholecystotomy, depending on the operative risk for the patient.

References
1. Greenfield LJ, Mulholland M, Oldham KT et al. Surgery scientific principals and practice. 2nd ed. Philadelphia: Lippincott-Raven, 1997.
2. Sabiston DC, Lyerly HK. Textbook of surgery. 15th ed. Philadelphia: W.B. Saunders Co., 1997.

Cholangitis
Alina D. Sholar

Background
Cholangitis is an acute bacterial infection of the biliary tree secondary to biliary obstruction. Originally described by Charcot in 1877, the mortality of acute toxic cholangitis is 20-25%.

Pathophysiology
Stones are the most common cause of obstruction, followed by malignancy. The most common bacterial isolates are:
- *E. coli*
- *Klebsiella*
- *Enterococcus*
- *Bacteroides*

As intrabiliary pressure rises, cholangiovenous reflux occurs at the level of the hepatic sinusoid, allowing bacterial translocation from the bile canaliculus into the vascular system, therefore causing the septic condition associated with cholangitis.

Charcot's triad consists of:
- Jaundice
- Fever
- Right upper quadrant pain

These describe the common clinical findings in cholangitis. Occasionally obtundation and hypotension (Reynold's pentad) indicate acute suppurative cholangitis.

Diagnosis and Radiologic Findings
Diagnosis by ultrasound identifies stones and ductal dilatation. Computed tomography is complementary to ultrasound and detects a dilated biliary tree, hepatic abscesses, and compressive masses. Cholangiography, either endoscopic (ERCP) or percutaneous (PTC) is used to evaluate the biliary tree, as well as to provide biliary drainage.

Management
Treatment is prompt intravenous fluid resuscitation, antibiotics directed against bacterial flora, and intensive observation of the patient. If a patient presents with septic shock, hemodynamic monitoring should be initiated. Patients who present with shock due to sepsis have an approximately 25% mortality rate. After initial resuscitation, transhepatic biliary drainage or endoscopic sphincterotomy and stone extraction or stent placement should be performed as indicated by the cause of the obstruction.

Emergency surgical decompression and common bile duct exploration are generally reserved for patients in whom PTC or ERCP has failed. After recovery from this illness, definitive surgery to correct the cause of obstruction can be performed (cholecystectomy for calculi, resection for malignancy, bile duct drainage procedure for strictures, etc.). In patients at high operative risk, palliation can be achieved with stenting of a tumor or stricture.

Primary Sclerosing Cholangitis

Background

Primary sclerosing cholangitis (PSC) is the development of multiple inflammatory strictures involving the intra- and extrahepatic biliary tree. With persistent obstruction, the disease is slowly progressive to cirrhosis and liver failure.

Epidemiology

Although no clear etiology is known, 70% of these patients have inflammatory bowel disease, usually ulcerative colitis. There is a male to female preponderance in a 2:1 ratio and at an average age of 40 years. Other rarely associated disorders are Reidel's thyroiditis, histiocytosis X, sarcoidosis, and retroperitoneal fibrosis.

The initial presentation of patients with sclerosing cholangitis is usually painless jaundice, pruritus, recurrent cholangitis, fatigue, and weight loss. Some may be asymptomatic with the disease only suspected by laboratory data.

Diagnosis

In nearly all patients, there is a two-fold elevation in alkaline phosphatase. Serologic evidence of autoimmune disease such as antinuclear antibody or antimitochondrial antibody is absent. Liver biopsy is performed for staging, identifying the extent of biliary cirrhosis.

Radiologic Findings

Diagnosis is confirmed by cholangiography, either percutaneously or endoscopically, featuring strictures alternating with dilated segments producing a characteristic beaded appearance of the bile ducts.

Management

Definitive treatment is liver transplantation for those with such severe cirrhosis as to cause liver failure. Endoscopic dilation of strictures to improve bile flow and relieve cholangitis is transiently palliative. Surgical excision of stenotic extrahepatic ducts with hepaticoenterostomy and transhepatic stenting in noncirrhotic patients is effective in alleviating jaundice, but requires frequent stent changes and makes later transplantation more difficult. Some studies suggest that ursodeoxycholic acid and methotrexate may be of some value. Cholangiocarcinoma develops or is present in 5-10% of patients with PSC; this is a contraindication to liver transplantation.

14

Primary Biliary Cirrhosis

Primary biliary cirrhosis is an autoimmune disease where cytotoxic T-cells attack biliary endothelium causing granulomatous destruction of intrahepatic bile ducts. This destruction leads to cholestasis, then progressive cirrhosis, portal hypertension, and liver failure. Patients often complain of fatigue and pruritus and jaundice is evident. This process is most prevalent in middle-aged women, especially in those with other preexisting autoimmune disorders. Laboratory evaluation elicits an elevated alkaline phosphatase level and positive antimitochondrial antibody. Hepatic transplantation is the only method of long-term treatment.

Choledocholithiasis

Alina D. Sholar

Background

Among patients undergoing cholecystectomy for calculi, 10-15% have common bile duct stones as well. Primary common duct stones will reoccur in 30% of patients within 5 years.

Pathophysiology

These stones are of two types: originating from the gallbladder (cholesterol stones) and primary common duct stones (calcium bilirubinate) associated with infection with bacteria or *Clonorchis sinensis*.

The most common complication of choledocholithiasis is obstructive jaundice. If the obstruction is unrelieved and bacteria are present, cholangitis can develop.

Diagnosis

Evaluation for preoperative detection of choledocholithiasis includes abnormal liver function tests. Elevation of alkaline phosphatase has the highest diagnostic accuracy, along with elevated total and direct bilirubin. Jaundice, acholic stools, and bilirubinuria are common presenting signs, although choledocholithiasis may also present as biliary colic, cholangitis, or pancreatitis.

Radiology Findings

Ultrasound shows the presence or absence of common duct stones only 30-50% of the time, however indirect signs such as cholelithiasis and dilatation of the biliary tree are useful in diagnosis. The normal common hepatic duct is 1 to 2.5 cm long and approximately 4 mm in diameter. The internal diameter of the normal common bile duct averages 5 mm.

Management

Preoperative management, when a dilated common bile duct (>8 mm) is seen or stones are detected on ultrasound, is ERCP with sphincterotomy and stone extraction. If this is successful and is free of complication, a laparoscopic cholecystectomy is performed in 24-48 hours. During the procedure, the common duct should be evaluated by intraoperative cholangiography. For patients who are debilitated or elderly, ERCP with sphincterotomy is sufficient and cholecystectomy does not have to be performed.

In up to 5% of patients undergoing cholecystectomy, common duct calculi are present without clinical abnormality to signal their preoperative detection. Use of intraoperative cholangiogram provides a road-map for the surgeon, identifies an anomalous cystic duct, and can detect ductal calculi. Many surgeons use intraoperative cholangiography selectively on only those patients with suspected duct stones or who have a dilated common duct while others use this routinely. Recall that 8% to 12% of patients with gall stones will have duct stones as well. Retrospective reviews have suggested that routine cholangiography has decreased negative explorations and overall morbidity of cholelithiasis.

14

When detected intraoperatively by cholangiography during laparoscopic chole-cystectomy, there are several ways to approach stone extraction:

1. Transcystic common duct exploration and stone extraction using baskets, balloons, and/or the choledochoscope is successful 90-95% of the time as determined by completion cholangiography.

2. Laparoscopic choledochotomy with common duct exploration and T-tube placement can also be done in experienced hands. If after a repeat T-tube cholangiogram at 10-14 days shows no further stones, the T-tube can be removed. If a stone is still present, the T-tube is left in situ, then removed in 6 weeks when a tract has formed and stones may be removed via the tract.

3. If neither of the above are performed due to either the surgeon's inexperience or reluctance to leave a T-tube for patient comfort reasons , a post-operative ERCP can be relied upon to remove stones. However this exposes the patient to the morbidities of the procedure, and reoperation would be required if the ERCP were to fail.

4. Open common bile duct exploration with T-tube placement is the traditional method of management.

5. Duodenotomy and sphincterotomy or sphincteroplasty is occasionally needed to remove impacted stones.

Choledochal Cysts

Background

Choledochal cysts are uncommon (incidence 1 in 100,000) and occur mostly in children under 10 yrs of age. Presentation is most commonly abdominal pain, a palpable right upper quadrant mass, or jaundice.

Anatomy

Type I- fusiform dilation of the common bile duct most common
Type II- diverticulum off of the common bile duct
Type III- cystic dilation of the intraduodenal bile duct, "choledochocele"
Type IV- combined extra- and intrahepatic cystic dilations of bile ducts
Type V- multiple intrahepatic biliary cystic dilation (Caroli's disease)

Radiology Findings

Radiological diagnosis is by ultrasound or computed tomography with definitive diagnosis provided by ERCP or PTC to demonstrate the size of the lesion, its extent, location, and type.

14

Management

Treatment is dependent on the type of bile duct cyst. Because of a 20-fold increased risk of cholangiocarcinoma in Type I cysts, surgical resection and Roux-en-Y hepaticojejunostomy anastomosis is the treatment. Cholecystectomy is also performed at this time. The treatment of Type II cysts is excision of the diverticulum. Type III is associated with a low risk of malignancy, so excision is not necessary. Endoscopic sphincterotomy may be performed if the patient develops occasional jaundice or cholangitis.

Caroli's Disease

Caroli's Disease is a congenital disease of the intrahepatic biliary tree characterized by multiple, saccular dilations of the bile ducts. Two types occur: the simple type and the periportal fibrosis type. The simple type is associated with recurrent cholangitis, liver abscesses, abdominal pain, and fever. It is also associated with "medullary sponge kidney," cystic lesions of the kidney, 60-80% of the time. The periportal fibrosis type has the same type of intrahepatic biliary dilation but also is associated with congenital hepatic fibrosis, cirrhosis, and portal hypertension. Cholangitis and kidney lesions are also commonly seen with this type. Caroli's disease is highly premalignant.

Presentation

Presentation is usually asymptomatic with the occasional presentation of abdominal pain, fever, chills, and mild jaundice. Advanced stages of Caroli's disease culminate with liver failure.

Diagnosis

Diagnosis is made by ultrasound, CT, or HIDA; definitive diagnosis relies on ERCP or PTC. If the patient becomes symptomatic, surgical intervention is warranted. If the disease is confined to a single lobe of the liver, excision of the affected lobe is indicated.

Management

If the patient is septic, temporary external drainage can be performed. In most cases, Roux-en-Y choledochojejunostomy with transhepatic stenting is necessary.

Prognosis

Unfortunately, the long-term prognosis is very poor with advanced disease and hepatic failure leading to the need for liver transplantation.

References

1. Greenfield LJ, Mulholland M, Oldham KT et al. Surgery Scientific Principals and Practice. 2nd ed. Philadelphia: Lippincott-Raven, 1997.
2. Sabiston DC, Lyerly HK. Textbook of Surgery. 15th ed. Philadelphia: W.B. Saunders Co., 1997

14

Common Bile Duct Injury

William R. Wrightson

Background

The rate of CBD injury is 0.2% and transiently increased with the introduction of laparoscopic cholecystectomy.

Etiology

Several factors are to blame.
- Mistaking the common duct for the cystic duct
- Distal clips placed on CBD

Table 14.1.   Classification of common bile duct injury

CBD Injury Classification

Type A	leaking cystic duct stump or from liver bed
Type B	injury to aberrant right hepatic duct
Type C	injury to right aberrant duct without ligation
Type D	lateral injury to the extrahepatic ducts (CBD)
Type E	circumferential injury to CBD

- Right hepatic duct mistaken for cystic duct
- Electrocautery injury
- Tear/laceration of duct

Presentation

Injuries discovered in the OR show bile staining or evidence of injury on the specimen or cholangiography. After a few days a patient may develop a biliary cutaneous fistula, bile peritonitis, jaundice and/or refractory abdominal pain.

Injuries discovered late are typically small injuries that have the end result of biliary stricture causing fever, fatigue and signs of cholangitis (Tables 14.1 and 14.2).

Management

The goals in management of biliary complications are:
1. Resuscitate
2. Define the anatomy
3. Provide drainage and function
4. Definitive repair.

CBD injury can be classified into various scales.

Diagnosis

Diagnosis can be made by several techniques. The optimal technique is ERCP. This provides both diagnosis of the problem as well as the potential for stenting and dilatation. Other options include:

- HIDA – sensitive but not specific
- T-tube sinogram
- PTC (stricture or transection)
- CT drainage of biliary ascites

14

Management

Choledochojejunostomy with Roux limb is the definitive procedure for an injury. This is done with an end-to-side single layer anastomosis and drained with T-tube or through the liver. A 60-70 cm length of jejunum is used and secured to the abdominal wall with an O ring (chimney) to allow for radiologic interventions.

Hepaticojejunostomy can be done if there is insufficient bile duct length. This is frequently the case. Remember that the bile duct blood supply is tenuous and the closer to the liver you take it the better chance for good supply. Intraparenchymal duct anastomosis may be necessary.

Table 14.2.   *Bismuth classification of bile duct injury*

Bismuth Classification

Level I	common duct with normal hepatic stump > 2 cm
Level II	hepatic duct stump < 2 cm
Level III	high stricture with preserved ductal confluence
Level IV	destruction of confluence
Level V	right sectorial duct with or without involvement of CBD

1. Left ductal system
   a. Segment II, III (Deep)
2. Right ductal system
   a. Segment V

References

1. Greenfield LJ, Mulholland M, Oldham KT et al. Surgery scientific principals and practice. 2nd ed. Philadelphia: Lippincott-Raven, 1997.
2. Sabiston DC, Lyerly HK. Textbook of surgery. 15th ed. Philadelphia: W.B. Saunders Co, 1997.

Cholangiocarcinoma and Gall Bladder Adenocarcinoma

Alina D. Sholar

Background

Cholangiocarcinoma is a rare occurrence and is seen in 0.01% to 0.5% of autopsies. This cancer has a dismal outcome with mortality approaching 80%-90% in 5 years.

Epidemiology

Bile duct cancers are almost always histologically adenocarcinoma, known as cholangiocarcinoma. Risk factors associated with the disease are: the presence of choledochal cysts and Caroli's disease, infestation with *Clonorchis sinensis,* hepatolithiasis, sclerosing cholangitis, ulcerative colitis, and exposure to Thorotrast.

The primary tumor may occur in three locations; one-third in the distal common bile duct, one-third in the common hepatic duct, and one-third in the right or left hepatic duct. When the tumor is located at the hepatic artery bifurcation, it is referred to as a Klatskin tumor.

Presentation

Painless jaundice is a common presentation of patients with bile duct cancer. If the biliary tree is obstructed at the level of the common bile duct, the gallbladder may be distended and palpable. This is known as Couvoisier's sign.

Diagnosis

Laboratory examination is compatible with obstructive jaundice with elevated bilirubin and alkaline phosphatase.

Radiology Findings

CT or ultrasound examination are the initial studies of choice to identify dilated bile ducts, the level of obstruction, and the presence of liver metastases. Further, more definitive imaging is achieved with percutaneous transhepatic cholangiography or endoscopic retrograde cholangiography. Brush biopsies can also be obtained at that time.

Management

Proximal tumors and Klatskin tumors are managed by resection of the extrahepatic biliary tree and bifurcation with or without partial hepatectomy, followed by bilateral hepaticojejunostomy. Lesions located in the middle of the common duct can be resected then choledochoenterostomy performed. For distal tumors, pancreaticoduodenectomy (Whipple procedure) is performed.

Nonresectability criteria include:

- Distant metastasis
- Extensive tumor extension into both lobes of the liver
- Involvement of the portal vein or main hepatic artery

Nonoperative palliation is achieved by placement of transhepatic stents either percutaneously or endoscopically. Surgical palliation for patients without disseminated tumor, but have unresectable hilar CA at exploration, is Roux-en-Y choledochojejunostomy with cholecystectomy.

Prognosis

Distal tumors have the most favorable prognosis with 5-year survival rates of about 40%. Lesions of the middle one-third of the bile duct have an approximate 5-year survival rate of 25%. Proximal tumors have a dismal prognosis of a 5-year rate of only 5%. Cholangitis is a common cause of death.

Gallbladder Cancer

Background

According to the National Cancer Institute's Surveillance, Epidemiology, and End-Results Program, gallbladder cancer is a very rare tumor, representing about 0.5% of all newly diagnosed carcinomas. It is the fifth most common cancer of GI tract. The average age of diagnosis is in the sixth and seventh decades of life and women are affected three times as often as men.

Epidemiology

Up to 70% of all patients with gallbladder cancer have gallstones. Some ethnic groups are at increased risk, including American Indians and Hispanics. Other risk factors are porcelain gallbladder, large gallstones >2.5 cm, and cholecystoenteric fistula. The incidence of cancer found incidentally at cholecystectomy is 1.0%.

14

About Pathophysiology

90% of gallbladder cancers are adenocarcinomas. These tumors spread by direct extension to the liver, regional metastasis to lymph nodes, and distant metastasis to the liver and peritoneal cavity.

Presentation

Some patients will likely present with symptoms related to their cholelithiasis, but because gallbladder cancer tends to spread by local extension, most patients' symptoms are related to invasion into surrounding structures at an advanced stage of the disease. Jaundice may develop from obstruction of the common bile duct or hepatic duct.

Diagnosis

Diagnosis is assisted by ultrasound and computed tomography. Most are found incidentally following an elective cholecystectomy.

Management

For management of disease limited to the mucosa, cholecystectomy is adequate. Tumor invading the muscularis or serosa requires cholecystectomy and wedge resection of a 2-3 cm liver bed margin adjacent to the gallbladder bed with lymph node dissection. If the tumor extends beyond the gallbladder, resection includes all of the previous, and additionally the adjacent common bile duct, then hepaticojejunostomy. Palliation for patients with unresectable disease includes systemic chemotherapy and regional radiotherapy with very low response rates. Choledochojejunostomy or hepaticojejunostomy can be utilized to relieve obstructive jaundice. If excessive operative risk is present, transhepatic or endoscopic stenting may be preferred.

Prognosis

Five-year survival is about 5% and median survival is 4-6 months.

References

1. Greenfield LJ, Mulholland M, Oldham KT et al. Surgery scientific principals and practice. 2nd ed. Philadelphia: Lippincott-Raven, 1997.
2. Sabiston DC, Lyerly HK. Textbook of surgery. 15th ed. Philadelphia: W.B. Saunders Co., 1997.

14

CHAPTER 15

Pancreas

William R. Wrightson

Clinical Focus on Pancreatic Disease and ERCP

Background

Acute pancreatitis is an inflammatory process that is manifest clinically by epigastric and back pain, nausea, and vomiting. In mild cases, these signs and symptoms gradually improve as the local inflammatory process resolves. In severe acute pancreatitis, systemic complications, including shock, renal failure, and adult respiratory distress syndrome, may supervene. The overall mortality of acute pancreatitis remains approximately 5% to 10%.

Etiology

Gallstones and alcohol are the two most common causes of acute pancreatitis, accounting for more than 80% of cases. A large number of less common causes (including medications, trauma, neoplasms, anatomic variants, metabolic problems) may be identified in 10% of patients.

Endoscopic Retrograde Cholangiopancreatography (ERCP)

Endoscopic retrograde cholangiopancreatography (ERCP) carries a substantial risk of postprocedure pancreatitis (overall > 5% in prospective trials but as high as > 20% in subgroups of patients). The experience at the University of Louisville is somewhat less at 1.5%. A large number of studies have evaluated various medications administered before ERCP in an effort to minimize the risk of post-ERCP pancreatitis, but to date most have been found to be ineffective. A relatively small trial of 62 patients, found that volume expansion with 10% Dextran-40 at a dose of 3 mL/kg before ERCP led to a decrease in serum amylase but did not significantly effect post-ERCP pancreatitis.

The Role of ERCP in the Management of Acute Pancreatitis

As the result of several randomized controlled trials, it is commonly accepted that ERCP is indicated in the acute setting when severe pancreatitis is complicated by progressive jaundice or cholangitis. Pancreatic ductal disruption or leak is a common event in severe pancreatitis, occurring in 37% of patients. Ductal disruption may lead to peripancreatic fluid collections, fistulae, pancreatic ascites, and necrosis and predicts a prolonged length of hospital stay.

Current Concepts in General Surgery: A Resident Review, edited by William R. Wrightson.
©2006 Landes Bioscience.

Sphincter of Oddi Dysfunction and Acute Idiopathic Pancreatitis

No obvious etiology is identifiable by history, laboratory tests, and standard imaging studies in 10% to 20% of patients with acute pancreatitis. Consensus has not been achieved to date on the optimal approach to the diagnostic evaluation of these patients. ERCP with sphincter of Oddi manometry in this patient population in may detect sphincter dysfunction and a possible cause of pancreatitis.

Symptoms improved in 83% of patients who were found to have elevated sphincter pressures compared to 33% of patients who had normal sphincter pressures. The complication rate of performing sphincterotomy in this patient population is real and is 20% to 40%.

Pancreatic Pseudocysts

Pancreatic pseudocysts are walled-off collections of pancreatic secretions that may form as a result of acute or chronic pancreatitis. Pancreatic pseudocyst is a common complication of chronic pancreatitis occurring in 20% to 40% of cases. Pseudocysts can be treated by endoscopic cystenterostomy or transpapillary drainage, percutaneously with computed tomography guidance or operatively.

Percutaneous and endoscopic drainage is most commonly performed via an indwelling catheter or stent that is left in place for several weeks, since earlier reports have shown a high recurrence rate after simple needle aspiration of pseudocysts.

Endoscopic treatment achieved complete resolution of the pseudocyst in 83%, and the other 17% eventually required surgery. Endoscopic drainage of pancreatic pseudocysts can be both safe and effective, and definitive treatment. It should be considered as an alternative option before standard surgical drainage in selected patients.

References

1. Ramdhaney S, Lapin, Vlodov J et al. Volume expansion in the prevention of severe post-ERCP pancreatitis. Am J Gastroenterol 2002; 97:S62. [Poster #536]
2. Mergener K, Baillie J. Endoscopic treatment for acute biliary pancreatitis: When and in whom? Gastroenterol Clin North Am 1999; 28:601-613.
3. Vitale GC, Lawhon JC, Larson GM et al. Endoscopic drainage of the pancreatic pseudocyst. Surgery Oct 1999; 126(4):616-621.

Thoracoscopic Splanchnicectomy for Pancreatic Pain

15

Parasympathetic efferent nerve signals travel to the upper gastrointestinal tract via the vagus nerves while sympathetic efferent messages descend through the splanchnic nerves. All pain sensations reach the central nervous system by the splanchnics. Division of these nerves is believed to provide relief from intractable pain of pancreatic origin from either cancer or pancreatitis.

Anatomy

The splanchnic nerves arise from the sixth through the ninth sympathetic gan-glia and descend on the lateral aspect of the vertebral bodies to exit the chest just behind the aorta. The lesser nerves stem from ganglia 10 and 11 and exit the chest about 1 cm posterior to the greater. Wide application of double lumen tube anes-thesia and modern videothoracoscopy have afforded a less invasive approach.

Thoracoscopic Splanchnicectomy

Operation may be done on only one side or bilaterally. The left side is divided for midline or left-sided pain. For predominantly right-sided pain the right is se-lected. The lateral position is preferred for a unilateral operation, prone for a bilat-eral procedure. After completion on one side with reexpansion of the lung the other side is done.

Two to three 5 mm ports are used. The chest is entered using the Optiview 5 mm trochar just below the scapula. A second port is used for the cautery or har-monic scapel. A third port may be necessary for retraction. CO_2 is insufflated into the chest obviating the need for split lung ventilation. At the conclusion of the case CO_2 is turned off, the patient is bag ventilated inducing a Valsalva and the last port is removed with the suction on. This results in minimal residual pneumothorax. In most cases this can be observed without chest tube placement as the CO_2 will readily be absorbed.

Results

The operation will be effective 85% of the time. In patients with chronic pancre-atitis the pain will often recur in a few months requiring a procedure on the con-tralateral side. Most patients with cancer will not need a second procedure. The operation has proven to be effective in one patient with duodenal carcinoma and in a patient with gastric carcinoma.

Selected Reading

1. Ihse I, Zoucas E, Gyllstedt E et al. Bilateral thoracoscopic splanchnicectomy: Ef-fects on pancreatic pain and function. Ann Surg 1999; 230:785-791.
2. Saenz A, Kuriansky J, Salvador L et al. Thoracoscopic splanchnicectomy for pain control in patients with unresectable carcinoma of the pancreas. Surg Endosc 2000; 14:717-720.

15

CHAPTER 16

Hernias

Timothy K. Bullock

Surgical History

In 1890, Bassini described a technique of repairing inguinal hernias and strengthening the inguinal floor. Shouldice later revised this concept in 1945. Plastic prosthesis were first introduced in 1954. In the 1960s Rives and Stoppa introduced the preperitoneal repair for unilateral and bilateral inguinal hernias respectively.[1] In 1986, Lichtenstein first described his tension free repair with mesh.[2] The 1990s saw the development of laparoscopic hernia repairs.[1]

Background

A hernia is an abnormal protrusion of a viscus or the peritoneum through a natural or aquired defect in the muscular wall of a cavity. The natural history of all hernias is progressive enlargement with possible incarceration and strangulation. Hernia repair is the most common surgical procedure with approximately 700,000 repairs done annually. Inguinal hernias account for 75% of all hernias. Two-thirds of these are indirect and one-third are direct. Ventral hernias comprise approximately 10% of hernias. Femoral hernias comprise only 3% of hernias and are much more common in women. Other hernia types include: umbilical hernias, epigastric hernia (defect in the linea alba), Richter's hernia (herniation of the antimesenteric border of a hollow viscus), Littre's hernia (sac occupied only by a Meckel's diverticulum), Spigelian hernia (defect in the linea semilunaris), obturator hernia, lumbar hernias, and other rare hernias.[3]

Etiology

All abdominal hernias occur through a defect in the transversalis fascia.[4] Indirect inguinal hernias are a result of a patent processus vaginalis. The processus vaginalis is formed by peritoneum that protruded through the deep inguinal ring during the descent of the testes in men and the course of the ovary and round ligament in women. Normally the processus vaginalis obliterates and failure to do so results in an indirect inguinal hernia.[3] Indirect hernias are more common on the right, presumably due to the later descent of the right testicle.[5]

Direct inguinal hernias are more common in the latter third of life. They occur through an attenuated transversus abdominus muscle and the transversalis fascia. Hernias through Hesselbach's triangle (bounded medially by the lateral border of the rectus sheath, laterally by the inferior epigastric vessels, and inferiorly by the inguinal ligament) are considered direct hernias. Indirect hernias occur lateral to Hasselbach's triangle.

Femoral hernias occur through the femoral canal (bounded anteriorly by the inguinal ligament, medially by the lacunar ligament, posteriorly by Cooper's ligament and laterally by the femoral vein). The female pelvis is broad and flat resulting in increased risk for femoral hernias.

Current Concepts in General Surgery: A Resident Review, edited by William R. Wrightson. ©2006 Landes Bioscience.

Pathophysiology

Abnormalities in collagen metabolism including: decreased hydroxylation, increased turnover of collagen molecules, increased serum elastase, and a decrease in antiproteolytic activity have been implicated as possible molecular causes of hernias. Patients with connective tissue disorders are at increased risk of hernias. Denervation of the abdominal wall has also been implicated as a risk factor, and an increased incidence of right-sided inguinal hernias is observed after appendectomy. Pregnancy appears to be a risk factor for femoral hernias. Obesity and strenuous physical activity have not been shown to increase the incidence of groin hernias.

Diagnosis

The diagnosis of a hernia is a clinical diagnosis based on the patient's symptoms and physical exam, rarely are any confirmatory studies required. Symptoms include vague discomfort, pain, fullness or a bulge. Extreme pain can occur with incarceration.

Physical examination includes inspection and palpation in both the supine and standing positions. A Valsalva maneuver or cough can also help in the detection of a hernia. Determination of whether a hernia is direct or indirect based on physical exam is usually unreliable. During examination an incarcerated hernia may be reduced with gentle pressure. Attempts at reduction should be abandoned if evidence of strangulation is present or if reduction would require undue force that might damage the contents of the hernia sac.

A femoral hernia presents as a bulge below the inguinal ligament. These hernias have a higher incidence of strangulation (22% at 3 months) than inguinal hernias (2.8% at 3 months). Incarcerated femoral hernias may also present with a bruit over the femoral vein.

Hydroceles can be differentiated from inguinal hernias by translumination. Only rarely are ultrasound, computed tomography, magnetic resonance imaging, or herniography needed to confirm the presence of a hernia.

An obturator hernia may present with a palpable mass on rectal or pelvic exam. One-half of these patients will present with pain in the hip extending down the medial thigh to the knee (Howship-Romberg sign).

Management

All hernias in adults should be repaired unless the patient's condition would result in an unacceptably high mortality. Support with trusses or surgical belts may be appropriate conservative management in the very high-risk patient.

Umbilical hernias in infants are very common. Umbilical hernias in adults, children over 4 years of age and large (>2 cm) umbilical hernias in children less than 4 years of age should be repaired.

An area of some controversy is the exploration of the contralateral groin in children with inguinal hernias. Presence of a hernia on the contralateral side may be as high as 50-75% in select groups of these patients. Most pediatric surgeons do explore the contralateral groin in boys under 2 years of age and girls under 4.

Operative Techniques

The layers of the abdominal wall from external to internal are as follows: skin, subcutaneous fat, Scarpa's fascia, Camper's fascia, external abdominal oblique, internal abdominal oblique, transversus abdominis, transversalis fascia, preperitoneal

16

fat, and the peritoneum. Numerous fascial layers condense, converge and insert forming important ligaments in the inguinal region.

The inguinal ligament is formed by the posterior reflection of the external abdominal oblique fascia. The inguinal canal is formed anteriorly by the aponeurosis of the external abdominal oblique, posteriorly by the transversus abdominis aponeurosis and the transversalis fascia, superiorly by the internal abdominal oblique and transversus abdominis muscles, and inferiorly by the inguinal and lacunar ligaments. The lacunar ligament is a triangular expansion of the inguinal ligament medially to insert on the pectin pubis.

The iliopubic tract is an aponeurotic band within the transversalis fascia. It attaches laterally to the ileopectineal arch and the anterior superior iliac spine. It travels medially to attach to the pubic tubercle and Cooper's ligament. Diagramatically, it is often confused with the inguinal ligament but is, in fact a separate stucture deep to the inguinal ligament. It is seen from within the preperitoneal space and is extremely important in preperitoneal and laparoscopic repairs.

Cooper's (pectineal) ligament is a strong fibrous band along the superior pubic ramus attaching medially to the lacunar ligament. The conjoint tendon is a convergence of aponeurosis of internal oblique and transverse abdominis muscles. It is present in less than 10% of cases.[3]

Table 16.1 describes the common inguinal hernia operations performed. Inguinal hernias can be repaired from an anterior, preperitoneal or intraperitoneal approach. They can be traditional or tension free repairs, laparoscopic or direct vision repairs, and they can be repaired with or without mesh. All open anterior inguinal hernia repairs share several features: exploration with exposure of necessary structures, reduction or ligation of the hernia sac, and closure of the abdominal wall defect with or without autogenous or mesh reinforcement. The type of repair performed is tailored to patient circumstances and the surgeon's preference. With the many options in hernia surgery it is important to balance the recurrence rate, the risk of complications, patient concerns, and the cost of the procedure.

Results and Complications

Complication rates for inguinal hernia repairs are reported in most studies to occur in 7-12% of patients. Common complications include hematoma, infection, and scrotal swelling. Less common, but more serious complications include: testicular atrophy, ischemic orchitis, leg swelling, pulmonary emboli, mesh infections, and paresthesias or numbness related to nerve entrapment or division.

Various studies report recurrence in 1 to 10% of primary inguinal hernia repairs and in 5-35% of recurrent inguinal hernia repairs. Mortality is reported in less than 1% of cases but can be as high as 13% in incarcerated or strangulated hernias.[8]

Laparoscopic repairs require knowledge of anatomy from within the pelvis. Misplaced staples can lead to unique complications. Staples should not be placed in the triangle of doom (ductus deferens medially and spermatic vessels laterally) since it contains the external iliac artery and vein as well as the femoral nerve. Special care should also be taken to avoid an accessory obturator artery that can cross Cooper's ligament. Staples should not be placed inferior to the iliopubic tract due to the risk of injury to the lateral femoral cutaneous or the femoral branch of the genitofemoral nerve.[3]

16

Table 16.1. Hernias

Name	Approach	Mesh vs. No Mesh	Tension vs. Tension Free	Method of Reinforcing Floor
Marcy repair	Open anterior	No mesh	Tension	No reinforcement of floor; internal ring narrowed with interrupted sutures[10]
Bassini repair	Open anterior	No mesh	Tension	Internal abdominal oblique, transverse abdominal, and transversalis fascia are sutured to the shelving edge of the inguinal ligament with interrupted sutures (Sabiston)
Shouldice repair	Open anterior	No mesh	Tension	Same layers as Bassini, but with running imbricating sutures
McVay (Cooper ligament) repair	Open anterior	No mesh	Tension	Internal abdominal oblique, transverse abdominal, and transversalis fascia are sutured to pubic tubercle and Cooper's ligament medially then transitioned to the inguinal ligament laterally
Lichtenstein repair	Open anterior	Mesh	Tension free	Mesh patch sutured to pubic tubercle and shelving edge of the inguinal ligament laterally and to the conjoined tendon medially
Plug and patch	Open anterior	Mesh	Tension free	Hernia sac reduced and plug of mesh placed in defect and sutured with onlay mesh patch usually not sutured
Stoppa or GPRVS	Open preperitoneal	Mesh (Mersilene)	Tension free	Large mesh placed over the myopectineal orifice (bordered by pubis inferiorly, rectus medially, iliopsoas laterally, and internal abdominal oblique and transversus abdominis superiorly). No sutures. Held by intra-abdominal pressure.[11]
TAPP	Laparoscopic transabdominal preperitoneal	Mesh	Tension free	Prosthetic mesh is tacked to consistent structures (Cooper's ligament, rectus muscle, transverse abdominis aponeurotic arch and the superior edge of the iliopubic tract).
TEPA preperitoneal	Laparoscopic	Mesh	Tension free	Prosthetic mesh is tacked to consistent structures (Cooper's ligament, rectus muscle, transverse abdominis aponeurotic arch and the superior edge of the iliopubic tract).
IPOM	Laparoscopic intraperitoneal	Mesh	Tension free	Prosthetic mesh is tacked to consistent structures (Cooper's ligament, rectus muscle, transverse abdominis aponeurotic arch and the superior edge of the iliopubic tract).

16

Other Hernias

Ventral hernias are the result of failed abdominal closure. Older age, obesity, pulmonary complications, abdominal distension, male gender, wound infections, type of incision and closure, and male gender have all been implicated as risk factors for incisional hernias. Incisional hernias occur in 2-10% of patients following laparotomy. Midline incisions may be at higher risk than transverse incisions for herniation. Wounds complicated by infection are 5 times as likely to be complicated by hernias. Primary incisional hernia repairs may recur in as many as 30-50% of repairs. Mesh repairs are reported to have recurrence rates of 10%. Small incisional hernias may be repaired primarily; however, mesh repairs are frequently indicated. Adequate repair can be performed open or laparoscopically. In patients who have lost the "right of abdominal domain," progressive pneumoperitoneum (sequential insufflation of air into the peritoneal cavity) may be useful.[9]

Umbilical hernias are usually congenital and usually close spontaneously by 2 years of age. Adult umbilical hernias are acquired. Their incidence is increased by factors that increase intra-abdominal pressure (ascites, pregnancy etc.). Repair can usually be performed primarily however mesh can be used for larger hernias.

Femoral hernias occur through the femoral canal medial to the femoral vein. These may present as a bulge below the inguinal ligament. Incarceration is common. Femoral hernias can be repaired with a McVay (Cooper's ligament) repair, preperitoneal repair, or laparoscopic repair.[3]

Relevant Studies

1. Robbins AW, Rutkow IM. The mesh-plug hernioplasty. Surg Clin North Am 1993; 73(3):501-12.
2. Rutkow IM, Robbins AW. Mesh plug hernia repair: a follow-up report. Surgery 1995; 117(5):597-8.
3. Lichtenstein IL, Shulman AG et al: The tension-free hernioplasty. Am J Surg 1989; 157:188.
4. Condon RE. Incision hernia. In: Nyhus LM, Condon RE, eds Hernia. 4th ed. Philadelphia: JB Lippincott, 1995.
5. Condon RE, Carilli S. The biology and anatomy of inguinofemoral hernia. Semin Laparosc Surg 1994; 1:75.

16

References

1. Holzheimer RG, Mannick JA. Surgical treatment: Evidence-based and problem-oriented. München zuckschwerdt 2001; 611-23.
2. Kurzer M, Belsham PA, Kark AE. The Lichtenstein repair. Surg Clin North Am 1998; 78(6):1025-46. Review.
3. Townsend CM, Sabiston DC. Sabiston textbook of surgery the biological basis of modern surgical practice. 16th ed. Philadelphia: W.B. Saunders Co., 2001.
4. Condon RE, Carilli S. The biology and anatomy of inguinofemoral hernia. Semin Laparosc Surg 1994; 1(2):75-85.
5. Schwartz SI. Principles of surgery. 7th ed. New York: McGraw-Hill, Health Professions Divisions, 1999.
6. Zona JZ. The incidence of positive contralateral inguinal exploration among preschool children—a retrospective and prospective study. J Pediatr Surg 1996; 31(5):656-60.
7. Wiener ES, Touloukian RJ, Rodgers BM et al. Hernia survey of the section on surgery of the American Academy of Pediatrics. J Pediatr Surg 1996; 31(8):1166-9.
8. MacFadyen Jr BV, Mathis CR. Inguinal herniorrhaphy: Complications and recurrences. Semin Laparosc Surg 1994; 1(2):128-140.
9. Santora TA, Roslyn JJ. Incisional hernia. Surg Clin North Am 1993; 73(3):557-70.
10. Griffith CA. The Marcy repair revisited. Surg Clin North Am 1984; 64(2):215-27.
11. Wantz GE. Giant prosthetic reinforcement of the visceral sac. The Stoppa groin hernia repair. Surg Clin North Am 1998; 78(6):1075-87. Review.

16

CHAPTER 17

Minimally Invasive Surgery (MIS)

Vincent C. Lusco III

Minimally Invasive Surgery and Physiologic Principals

Introduction

The field of minimally invasive surgery (MIS) has applications in all surgical disciplines.

Although this is a relatively new field, it is one of the fastest developing, and patient demands for more and more procedures performed through smaller incisions continues to grow.

Applications of the principles of MIS can be found in the following:

- Video-assisted thoracoscopic procedures
- Transplant surgery and the advent of laparoscopic donor nephrectomies which has increased the donor renal pool.
- Sentinel lymph node technology and radioguided parathyroid surgery.
- Gynecologic procedures include endoscopic techniques such as hysteroscopy to laparoscopic-assisted vaginal hysterectomies.
- Urologic surgeons perform cystoscopy, ureteroscopy and radical prostatectomies through a laparoscopic approach.
- Bariatric surgery with laparoscopic Roux-en-Y bypass and gastric banding procedures.

The goal of this chapter is to discuss the history, principles and common procedures performed by general surgeons.

The physiology and complications that are associated with laparoscopy will also be explored.

Surgical History

In the 1880s, the cystoscope was perfected by Maximillian Nitze. Twenty years later, Georg Kelling of Dresden placed a cystoscope into the inflated abdomen of a dog.

The term laparoscopy was coined by Hans Jacobaeus, and he was the first to publish about thoracoscopy and laparoscopy in humans.

Erich Mühe performed the first laparoscopic cholecystectomy in 1985.

There was a major technological advance in the mid 1980s when the high resolution charge couple device was developed for video cameras.

This led to popularization of the laparoscopic cholecystectomy by Phillippe Mouret in 1987.

The versatility of laparoscopy was demonstrated when the first laparoscopic appendectomy was performed by Semm in 1983, and the first laparoscopic nissen fundoplication was carried out by Geagea in 1991.

Current Concepts in General Surgery: A Resident Review, edited by William R. Wrightson.
©2006 Landes Bioscience.

Principles

Overall advantages with laparoscopic procedures
- Decreased postoperative pain
- Shorter duration of postoperative ileus
- Earlier discharge
- More rapid recovery
- Patient satisfaction
- Improved cosmesis

As noted by Sir Alfred Cusheri, the advantage of laparoscopic procedures is directly related to incisional trauma and those proceedures where the incision is the most traumatic event in the case will have the greater advantage.

The ideal gas for laparoscopy is one that is inert, water-soluble, and noncombustible.

Although no gas meets all of these criteria, CO_2 is commonly used, because it is water-soluble and noncombustible.

CO_2 is not inert, because it is rapidly absorbed and alters many physiologic parameters (Table 17.1).

Economics

With laparoscopy there is a significant capital investment.

This coupled with longer operating times and disposable instruments increases the costs of a procedure dramatically.

These costs however, can be offset by a decreased length of stay, shorter convalescent period, and a more rapid return to work.

This has been demonstrated in laparoscopic cholecystectomy, fundoplication, splenectomy and adrenalectomy.

Less cost benefit is seen with inguinal herniorrhaphy since this is already an outpatient procedure.

Physiology of Pneumoperitoneum

Hemodynamic Changes

Pneumoperitoneum results in an overall decrease in preload, accompanied by an increase in afterload.

There is an increase in central venous pressure and pulmonary capillary wedge pressure with a paradoxical decrease in cardiac chamber filling.

Systemic vascular resistance and mean arterial pressure increase secondary to release of vasopressin and catecholamines from direct aortic compression by the insufflated gas.

Table 17.1. Potential gases for pneumoperitoneum and their characteristics

Gases	CO_2	Air	NO_2	He	O_2
Combustible	No	Yes	Yes	No	Yes
Water soluble	Yes	No	No	No	No
Inert	No	Yes	Yes	Yes	No
Peritoneal irritation	Yes	Yes	No	No	Yes

17

Laparoscopy can also be associated with bradycardia or sinus arrest secondary to vagal stimulation from peritoneal stretching. There is also a decrease in mesenteric blood flow.

Changes in Pulmonary Function

Ventilation can be more difficult during laparoscopy due to the cephalad deflection of the diaphragm. The work of breathing is characterized by increasing pulmonary arterial pressures and decreasing tidal volumes, functional residual capacity, and compliance.

The ventilatory rate must be adjusted to avoid systemic acidosis and hypercapnia.

Healthy subjects have no change in their pCO_2 with 10-15 mm Hg of intraabdominal pressure. The benefit of laparoscopy is less pulmonary embarrassment postoperatively when compared to open procedures. This is demonstrated by improved FVC, FEV1 and peak expiratory flow rates postoperatively. In summary, pulmonary changes are usually well tolerated unless the patient has chronic pulmonary disease in which close monitoring is necessary.

Changes in Immune Function

This is a controversial topic with regards to oncologic laparoscopy that is under study at this time.

Although improved systemic cell-mediated immunity after laparoscopic procedures has been documented, impairment of intraperitoneal cell-mediated immunity has been shown with pneumoperitoneum.

The mechanism of this is unknown, but in vitro studies have shown macrophages incubated in CO_2 produced significantly less tumor necrosis factor and IL-1 in response to LPS compared to controls.

Others have implicated a stimulatory effect on tumors by CO_2.

This combined with excessive manipulation of the tumor and difficulty in obtaining negative margins or decreased lymph node sampling has led to criticism in the use of laparoscopy for oncologic procedures.

However, clinical studies comparing open to laparoscopic colon resections have not shown an increased rate of tumor recurrence at follow-up of 5 years.

Risk of Thromboembolism

Compression of the IVC from elevated intraabdominal pressure leads to sluggish blood flow through the femoral vessels. This combined with a head up position can lead to thrombus formation. Current recommendations are TEDS/SCDs with or without LMWH for longer cases or patients at higher risk.

Fluid Balance

Several factors lead to decreased fluid loss or ambiguity of volume status during laparoscopy.

- Evaporative fluid loss normally seen with open procedures is negligible. In addition, insufflated air is usually humidified.
- Increased intraabdominal pressure results in decreased renal blood flow with resultant decrease in GFR and urine output. This is aggravated by a secondary release of renin and ADH resulting in further reabsorption of sodium and free water leading to anuria.

The transient decrease in intraoperative blood flow however, does not seem to produce significant renal dysfunction postoperatively.

Laparoscopy during Pregnancy

The two most common surgical procedures performed during the gestational period are for appendicitis and gallbladder disease (acute cholecystitis and symptomatic cholelithiasis), both of which are completely safe to perform during the second trimester.

- First trimester—generally contraindicated unless emergency because of 12% miscarriage rate and risk of teratogenesis
- Second trimester—regarded as the safest time. The risk of miscarriage and teratogenesis is close to 0% and the rate of preterm labor is 5-8%
- Third trimester—difficult due to inadequate visualization and 30% risk of preterm labor.

Precautions for laparoscopy in pregnant patients include, lower intraabdominal pressures, minimizing the degree of reverse Trendelenburg and manipulation of uterus, and placement of the patient in the left lateral position to avoid compression of the vena cava.

Complications

In addition to complications of pneumoperitoneum discussed above, numerous complications are associated with trocar and veress needle placement.

Other procedure-specific complications will be discussed later, for example common bile duct injury with laparoscopic cholecystectomy.

Visceral

- Generally occurs during placement of trocars.
- Small bowel is the most commonly injured, followed by colon.
- Injury to bowel is associated with a 5% mortality and missed injuries can present with delayed sepsis.
- Bladder injuries are more frequent when a Foley is not placed preoperatively and can be detected by gas in the Foley bag or hematuria.

Vascular Injuries

- Rare
- Seen mostly with pelvic procedures
- Most injuries involve distal aorta or IVC and branches.

Wound Infections

- Overall rate of 0.1-3%. More commonly located at trocar site where inflamed gallbladder or appendix is removed.

Hernias

- 5 mm port sites do not need repair.
- 10 mm port sites below the umbilicus should have fascial repair.
- 10 mm port sites above the umbilicus are repaired only if stretched or dilated.
- More common in the midline where the abdominal wall is thinnest.

17

Venous Gas Embolism

- Occurs with venous injury, usually during access, which allows insufflated air to enter venous system.
- Typical scenario is acute hypotension, cyanosis and hypoxia with a "mill wheel" murmur.
- Treatment is placement in Trendelenburg and left lateral decubitus position and aspiration of gas from right ventricle through a central line.

Selected Reading

1. Brook D. Current review of minimally invasive surgery. Springer-Verlag, 1998.
2. Corson J, Williamson R. Surgery. Mosby: 2001:13.1-13.12.
3. Eubanks W, Swanstrom L, Soper N. Mastery of endoscopic and laparoscopic surgery. Lippincott Williams & Wilkins, 2000.
4. Macfaylyen Jr B, Litwin D, Park A. Laparoscopic splenectomy, part I, indications and outcomes. Contemp Surg 2000; 56(6):345-354.
5. Rosen M, Ponsky J. Minimally invasive surgery. Endoscopy 2001; 33(4):358-366.
6. Sabiston D. Textbook of surgery. 16th ed. W.B. Saunders Company, 2001:292-310.
7. Schwartz S. Principles of surgery. 7th ed. McGraw-Hill, 1999:2145-2162.

Common Laparoscopic Procedures and Indications

Cholecystectomy

Laparoscopic cholecystectomy has become the standard of care for symptomatic cholelithiasis, acute and chronic cholecystitis, and biliary dyskinesia (defined as an ejection fraction < 35% on HIDA scan). There is a relatively low morbidity 1-9% with conversion to open rates of 1.8-7.8%. This procedure epitomizes the principles of laparoscopy as outlined earlier. The major complication seen with laparoscopic cholecystectomy is injury to the common bile duct (CBD), which occurs 0.2-0.7%. **This is more frequent than with the open procedure**. Techniques to reduce injuries to the CBD include use of a 30° scope with cephalic and lateral retraction of the gallbladder to expose the triangle of Calot. Other complications include hemorrhage, bile leaks, retained stones and pancreatitis.

Indications for intraoperative cholangiogram, include elevated liver function tests, dilated common bile duct, ambiguous anatomy, inability to clear the CBD by preoperative ERCP, and the presence of multiple small stones.

Intraoperative strategies for common bile duct stones:

- Stones < 3 mm-IV glucagon and rapid infusion of saline through cholangiocath into CBD.
- > 3 mm, or if CBD dilated-best removed through cystic duct with a stone basket.
- > 8 mm- choledochotomy, stone extraction and T-tube.

Laparoscopic CBD exploration has been shown to be as effective and more cost-efficient as postop ERCP.

17

Anti-Reflux Procedures

Laparoscopic Nissen fundoplication has become the accepted procedure for treatment of gastroesophageal reflux disease and the second most common laparoscopic abdominal procedure performed today. For patients with achalasia, laparoscopic longitudinal (Heller) myotomy has also been shown to be effective and feasible.

Ideal Patient
- Typical symptoms, heartburn and regurgitation, associated with stricture or Barrett's esophagus.
- pH proven pathologic acid exposure.
- Dependent on proton pump inhibitors for symptomatic relief.
- Respiratory symptoms associated with GERD.

Workup
- EGD to evaluate for erosive esophagitis, strictures or Barrett's esophagitis
- 24 hour pH monitoring
- Video esophagogram to evaluate for esophageal length (shortening seen with strictures) or large hiatal hernias > 5 cm
- Gastric emptying study should be done in patients with longstanding diabetes, vomiting or peptic ulcer disease.

Complications
- Rate 8%
- 2% require conversion to open procedure
- Pneumothorax/mediastinum due to breach of pleural membranes during hiatal dissection, does not usually require a chest tube.
- Bleeding from short gastrics, spleen or liver retraction

Splenectomy

Indications
The most common indication is for ITP when glucocorticoid therapy fails to result in adequate platelet count. Other indications include spherocytosis, ellitocytosis, Hodgkin's and nonHodgkin's lymphomas, autoimmune hemolytic anemia, leukemia with hypersplenism, splenic cysts and nonmalignant tumors.

Contraindications
Splenomegaly (18-20 cm), cirrhosis with portal hypertension secondary to presence of varices, and coagulopathy.

Controversies
Splenic artery embolization preop:
- Pros: reduce blood loss, increase platelet count, decrease splenic size by 10-20%.
- Cons: presurgical pain, migration of coils to liver, abscesses, pancreatitis, can obscure the dissection planes across the hilum.

17

Complications

Overlooking accessory spleens, iatrogenic pancreatic, colon or gastric injury, intra/postop hemorrhage, left pleural effusion/atelectasis, or pneumonia.

Inguinal Herniorrhapy

Most controversial laparoscopic procedure
- Does not result in shorter hospital stay
- Long term efficacy pending
- Increased OR costs
- Requires general anesthesia and potential risk for injury to abdominal organs

Types

Transabdominal Preperitoneal (TAPP) Repair
- Most commonly performed with easier identification of anatomy
- Increased risk of visceral injury.

Totally Extraperitoneal Repair (TEP)
- More difficult
- Ideal for patients with history of lower abdominal surgery
- Leading indication—recurrent hernias
- Amenable for bilateral hernia repair
- Contraindicated in setting of incarcerated hernia

Adrenalectomy

Most adrenal neoplasms are amenable to laparoscopic approach because of small size and benign nature. The right adrenal gland is pyramidal in shape and lies superior to the right kidney. Laparoscopic right adrenalectomy is easier than left adrenalectomy but can be more hazardous due to shorter adrenal vein that empties into the vena cava. The left adrenal gland is more flattened and in intimate contact with the medial aspect of the superior pole of left kidney. The key to laparoscopic resection is complete division of the splenorenal ligament superiorly to the diaphragm and mobilization of the spleen medially.

Indications
- Aldosteronoma (Conn's syndrome)
- Cushing's syndrome (cortisol-secreting adrenal adenoma, primary adrenal hyperplasia)
- Pheochromocytoma (sporadic or familial)
- Nonfunctioning cortical adenoma, > 4 cm or suspicious radiographic appearance
- Adrenal mets

Contraindications
- Adrenocortical carcinoma
- Malignant pheochromocytoma
- Large benign adrenal mass > 8-10 cm

17

References

1. Brook D. Current review of minimally invasive surgery. Springer-Verlag, 1998.
2. Corson J, Williamson R. Surgery. Mosby 2001: 13(12):1-13.
3. Eubanks W, Swanstrom L, Soper N. Mastery of endoscopic and laparoscopic surgery. Lippincott Williams & Wilkins 2000.
4. Macfaylyen Jr B, Litwin D, Park A et al. Laparoscopic splenectomy, Part I, indications, and outcomes. Contemp Surg 2000; 56 (6):345-354.
5. Rosen M, Ponsky J. Minimally invasive surgery endoscopy. 2001; 33 (4):358-366.
6. Sabiston D. Textbook of surgery. 16th ed. W.B. Saunders Company, 2001:292-310.
7. Schwartz S. Principles of surgery. 7th Edition. McGraw-Hill, 1999:2145-2162.

CHAPTER 18

Vascular Surgery

William R. Wrightson

Carotid Endarterectomy

Background

Stroke is a leading cause of morbidity and mortality in the US with 500,000 cases each year and 200,000 of these dying as a result. Carotid endarterectomy (CEA) serves to reduce the risk of stroke and has been found to do this better than medical management alone. The first CEA was performed by DeBakey in 1953 and since that time, the utility of carotid endarterectomy (CEA) has been well documented by a series of randomized controlled studies. The North American Symptomatic Carotid Endarterectomy Trial (NASCET) is probably the most significant carotid study to date. It showed a decrease in stroke rate from 26% to 9% with CEA compared to medical management alone (Table 18.1).

Presentation

Patients will have varied symptoms based on the distribution of their disease. In general, patients will present with a neurologic deficit that can be traced to the carotid circulation. Weakness, dizziness and drop attacks suggest a global event or vertebrobasilar disease and not carotid stenosis.

Transient ischemic attack (TIA) – neurologic event that reverses completely in < 24 hours. These may include amaurosis fugax (reversible retinal stroke).

Reversable ischemic neurologic deficit (RIND) – neurologic deficit that resolves completely within 3 weeks.

Stroke – Residual neurologic deficit that remains with time.

These forms of stroke are the result of embolic events to the cerebral blood circulation.

Pathophysiology

The development of carotid stenosis is secondary to atherosclerosis. Critical stenosis results when the luminal diameter in decreased by 50% to 70% which corresponds to a decrease in cross sectional area by 75% to 90%.

Diagnosis

Evaluation of the symptomatic patient consists of duplex scanning. Peak systolic velocities (PSV) in excess of 130 cm/s to 325 cm/s suggest significant stenosis. Huston et al report that PSV > 230 cm/s has a sensitivity of 86%, sensitivity of 90% and positive predictive value of 82% of identifying stenosis > 70%. An end diastolic velocity of > 70 cm/s suggests stenosis > 70 (Table 18.2).[1]

Angiography as discussed in earlier has inherent risk of stroke (1%) and does not offer and advantage in the diagnosis. This is a topic of discussion with recent reports

Table 18.1. Stroke risk with surgery or medical management

Study	Degree Stenosis	Follow-Up	CEA	Medical
Symptomatic				
North American Symptomatic Carotid Endarterectomy Trial (NASCET) stroke rate at 2 years	> 70%	2 years	9%	26%
North American Symptomatic Carotid Endarterectomy Trial (NASCET) stroke rate at 5 years	50-70%	5 years	15%	22%
Asymptomatic				
Asymptomatic Carotid Atherosclerosis Study (ACAS)	> 60%	5 years	5%	11%

suggesting that there is a poor correlation with the degree of stenosis in the symptomatic patient. Current recommendations are duplex scanning and angiography if post-stenosis velocities do not return to normal.

Management

Carotid endarterectomy is done through a neck incision along the anterior border of the sternocleidomastoid muscle. The carotid is isolated with vessel loops. The patient is heparinized to weight and the internal carotid is then clamped after 3 minutes. Some surgeons routinely shunt at this point regardless of backbleed pressures (> 50 mm Hg). Current recommendations are to shunt if expected clamp time is > 20 minutes and/or poor retrograde pressure. Injection of the carotid body with lidocaine is an option for control of pressure shifts due to manipulation.

Current studies recommend closure with some type of patch angioplasty to minimize restenosis. The optimal choice of material has been left to surgeon choice. Allen et al report their experience with ePTFE vs. saphenous vein patch and concluded that postoperative complications were the same but vein harvest site complications are eliminated with the use of ePTFE.[2]

Post procedure duplex scanning to assess for thrombosis, intimal flap or stenosis are more widely done. Approximately 5-8% of cases that undergo intraoperative duplex scanning are found to have stenosis that warrant revision.[3,4]

Table 18.2. Duplex velocities and degree of stenosis

	Sensitivity (%)	Specificity (%)	Positive Predictive Value (%)
Stenosis > 50%			
PSV > 130 cm/s	92	89	90
EDV > 40 cm/s	88	85	86
Stenosis > 70%			
PSV > 230 cm/s	86	90	83
EDV > 70 cm/s	82	89	81

18

References

1. Huston J, Meredith JE, Brown RD et al. Redefined duplex ultrasonographic criteria for diagnosis of carotid artery stenosis. Mayo Clinic Proceedings Nov 2000; V75(11):1133-1140.
2. Allen P, Jackson MR, O'Donnell SD et al. Spahenous vein versus polytetrafluoroethylene carotid patch angioplasty. Am J Surg 1997; V174(2):115-117.
3. Mays BW, Towne JB, Seabrook GR et al. Inraoperative carotid evaluation. Arch Surg 2000; V135(5):525-529.
4. Padayachee TS, Brooks MD, McGuinness CL et al. Value of intraoperative duplex imaging during supervised carotid endarterectomy. Br J Surg 2001; V88(3):389-392.

Abdominal Aortic Aneurysms

Background

In 1951, Dubost performed the first successful aortic aneurysm repair in Paris. Abdominal aortic aneurysms (AAA) are responsible for approximately 15,000 deaths annually. Approximately 3-4% of adults older than 65 have evidence of AAA. Surgery has become the treatment of choice; however, management of recognized aneurysms is ultimately dependent on size, rate of expansion and patient co-morbidities. Currently there are approximately 80,000 aortic repairs done annually.

Epidemiology

The incidence of aortic aneurysmal disease in the general population is 2% but increases to 20% in those with peripheral vascular disease. Risk factors for the development of AAAs include family history, male gender, age > 65, smoking, and hypertension.

Anatomy

The normal caliber of the abdominal aorta is 2 cm and runs from the aortic hiatus (T12) to the aortic bifurcation (L4).

Pathophysiology

AAA are believed to be atherosclerotic in origin. Several factors have been suggested including genetic alterations in biochemical structure of the aortic wall as well as infectious and immune factors. One feature of AAA is the breakdown of the extracellular matrix of the aortic wall. The primary deficiency has been found to be elastin. This may be due to deficiencies in formation or increased breakdown by proteases such as elastin. Marfan's syndrome patients, prone to AAA, have a defect in the fibrillin gene and have a deficiency in functional fibrillin. Fibrillin is a glycoprotein that acts as the scaffolding for elastin.

Presentation

Most patients are asymptomatic and are identified on physical exam or incidentally on imaging studies. Symptomatic patients are those with early leak or rupture and are a surgical emergency. The classic triad of back/abdominal pain, hypotension and pulsatile abdominal mass suggest rupture and the need for emergent surgery.

Diagnosis

Mass screening for AAA is not feasible from a cost perspective and is therefore limited to high risk patients.

18

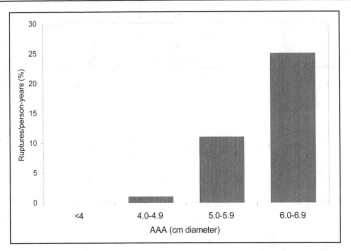

Figure 18.1. Risk of aortic aneurysm rupture based of size. Adapted from Hallett J. Management of abdominal aortic aneurysms. Mayo Clinic Proceedings April 2000; Vol. 75(4): 395-399.

Ultrasound has a sensitivity of 99-100% and can easily be used to longitudinally follow AAA. Ultrasound is operator-dependent but has been found to be reproducible to within 0.3 cm.

CT scan is 100% sensitive and specific and is probably more accurate than US. CT is costly and remains secondary for following aneurysms over time.

Rupture

The risk of rupture of AAA is related to size and rate of expansion. There have been multiple studies that have found a consistent relationship with risk of rupture and size > 5 cm. Expansion rates vary as well with the highest risk in those expanding at a rate > 0.4 cm/year. The most important factor affecting this rate, based on the Law of Laplace, is blood pressure. The risk of rupture for aneurysms < 4 cm is 2%, 4-5 cm 3-12% and > 5 cm 25-41%. The figure demonstrated rupture risk in person years.

Only 50% of patients with rupture have the classic signs. However any patient with hypotension and abdominal pain with a known history of AAA is taken for surgery. Aneurysms typically rupture into the retroperitoneum and are frequently tamponaded initially. Resuscitation to a euvolemic state should be tempered and take place in the operating room if the patient is conscious and has adequate peripheral perfusion (Fig. 18.1).

Management

Medical management involved control of blood pressure and elimination of risk factors such as smoking. Current trials have investigated the use of tetracycline and macrolide antibiotics to control the expansion of AAAs. Use of propranolol and anti-inflammatory agents has been used as well.

18

Coronary artery disease (CAD) is present in 50% of patients with AAA. CAD remains the leading perioperative cause of death after operations of the abdominal aorta. The incidence of fatal MI has been reported to be as high as 5% and nonfatal MI as high as 16%. Thorough cardiac evaluation is required in patients planned for elective AAA repair.

Current recommendations are for elective repair of aneurysms > 5 cm. The Small Aneurysm Trial demonstrated no long term survival advantage with surgery over ultrasound surveillance unless the aneurysm exceeded 5.5 cm. A report from the Society of Vascular Surgery suggested repair of asymptomatic aneyrysms that are twice the normal diameter of the infrarenal aorta provided there are no contraindications to surgery. If elective repair is contraindicated patients with AAA < 5 cm should be followed with US every 6 months and every 3 months for AAA > 5 cm. Expansion > 0.5 cm is an indication for surgery. Contraindications for surgery include:

- Myocardial infarction within the past 6 months
- Chronic renal failure
- Stroke
- COPD with dypsnea at rest
- Life expectancy < 2 years

Surgical repair can be through a transperitoneal or retroperitoneal approach. Both offer advantages and are tailored to the clinical circumstance. Retroperitoneal repair is associated with fewer postoperative complications, short stay in the hospital and intensive care unit.[4] The transperitoneal approach is advantageous for concomitant intraperitonal disease treatment, extension to the right iliac artery and ruptured AAA. The AAA is typically repaired with a Dacron or polytetrafluoroethylene tube or bifurcated graft with the aneurysmal sac closed over the graft.

Complications

The most common complications following surgery are cardiac events, pulmonary insufficiency, acute renal insufficiency, thromboembolization and wound infection. Ischemic colitis is a known rare complication and is the result of ligation of the inferior mesenteric artery. Patients with heme positive stools or evidence of acidosis postoperatively should be evaluated with colonoscopy.

References

1. Hallett J. Management of abdominal aortic aneurysms. Mayo Clinic Proceedings April 2000; 75(4):395-399.
2. Greenfield LJ, Mulholland M, Oldham KT et al. Surgery scientific principals and practice. 2nd ed. Philadelphia: Lippincott-Raven, 1997.
3. Vammen S, Lindholt JS, Ostergaard L et al. Randomized double-blind controlled trial of roxithromycin for prevention of abdominal aortic aneurysm expansion. Br J Surg April 2001; 88(8):1066-72.
4. Jing Z, Cao G, Ye B. A comparative study on transabdominal versus retroperitoneal approach for abdominal aortic surgery. Zhonghua Wai Ke Za Zhi Jan 1998; 36(1):20-2.

Endovascular and Noninvasive Vascular Surgery

Background

Seldinger described the needle-wire catheter technique in 1953 and signaled the beginning of a new era in vascular and endovascular therapy. Parodi and Palmaz performed the first endovascular repair of an infrarenal AAA in 1991. Minimally invasive surgical techniques have potential applications in a substantial part of the population with vascular disease. An increasing aged population with significant comorbidities, make endovascular approaches appealing. Elective open repair remains a relatively safe procedure with mortality from 0-4%; however avoidance of a large abdominal operation and the associated complications would be advantageous.

Abdominal Aortic Aneurysm Repair

Current Criteria

1. Asymptomatic infrarenal or common iliac aneurysm
2. Anatomy suitable for repair
3. Aneurysm neck length > 10-15 mm
4. External and common iliac arteries must accommodate the device (> 8 mm)

Technique

Procedures are performed in a dedicated endovascular suite with operative capabilities. Angiographic instrumentation includes a fixed angiographic table, fluoroscopy with digital subtraction and magnification and endovascular ultrasound. Patients are usually done under regional or general anesthesia. Local has been done in some instances.

Bilateral femoral artery cut downs are performed with arterotomy. A 20-27°F sheath is introduced to facilitate transfer of the device. The endovascular graft is introduced over a wire and deployed with fluoroscopic and ultrasonic assistance. Bifurcated grafts are deployed on one side initially with the second limb deployed and tied in subsequently.

Tube Graft

Endovascular repair has several options in terms of type of repair used. Early investigations utilized a single tube graft for isolated infrarenal aortic lesions. Others have used an aortoiliac graft with femoral-to-femoral bypass. The second and third generation devices allow for aortoiliac repair. The significance of continued perfusion of the aneurysm sac by lumber and IMA vessels is unclear. A study of 831 patients who underwent an aortic aneurysm exclusion found 2% with continued perfusion.

Outcomes

The application of endovascular AAA repair is comparable in terms of morbidity and mortality to open techniques. This technique remains under scrutiny. Ohki et al report a significant number of complications following repair.[1] These include endovascular leaks, rupture, and failed grafts. Many of these could be corrected with noninvasive techniques or open procedure (10%). In most instances the technical success rate ranges from 85-95%.

18

Angioplasty and Stenting

Transluminal dilatation was first described by Dotter in 1964. He subsequently introduced balloon dilatation (1965) and stents (1969). Current strategies include balloon dilatation of larger vessels with or without stenting. This can resolve short segment stenosis and, for example, enhance inflow to the lower extremities for outflow disease.

Access

Vascular access is usually through the femoral artery by percutaneous cannulation. A sheath is then placed and various guide wires are used to traverse the lesion.

- Standard solid wire with outer spiral
- Teflon coated (decrease friction)
- J wire (allow extension and retraction of J)
- Floppy tipped guide wire
- Steerable wires
- Hydrophilic wires (decreased friction)

Balloon Dilatation

The balloon is placed across the lesion and inflated. This results in fracturing of the plaque enlarging the lumen. Gradients > 20 mm Hg across the lesion suggest inadequate dilatation. There is a 90% initial success rate for short segment focal lesions and a 4 year patency rate of 60-70%.

Stenting

Stents are typically used in vessels with large diameters and high flow rates. The basic stent designs are usually stainless steel spring loaded, thermally expanded memory stents and balloon expandable stents.

References

1. Patterson MA, Jean-Claude JM, Crain MR et al. Lessons learned in adopting endovascular techniques for treating abdominal aortic aneurysms. Arch Surg 2001; 136:627-634.
2. Ohki T, Veith FJ, Shaw P et al. Increasing incidence of midterm and long term complications after endovascular graft repair of abdominal aortic aneurysms: A note of caution based on a 9-year experience. Ann Surg 2001; 234:323-335.
3. Sicard GA, Rubin BG, Sanchez LA et al. Endoluminal graft repair for abdominal aortic aneurysms in high risk patients and octogenarians. Ann Surg 234:427-437.
4. May J, Woodburn K, White G. Endovascular treatment of infrarenal abdominal aortic aneurysms. Ann Vasc Surg 1998; 12:391-395.
5. Hallett JW, Marshall DM, Patterson TM. Graft related complications after abdominal aortic aneurysm repair: Reassurance from a 36 year population based experience. J Vasc Surg 1997; 25:277-284.

CHAPTER 19

Cardiothoracic Surgery

William R. Wrightson

Lung Cancer

Surgical History

Milton Anthony performed the first successful thoracotomy with lung resection in 1821. The first pneumonectomy was performed by Sir William Macewen in 1895 with a later single stage pneumonectomy by Evarts Graham in 1933 for squamous cell carcinoma.

Background

There are over 170,000 new cases of lung cancer reported each year with an expected mortality of 100,000 for men and 70,000 for women. The overall prognosis for lung cancer is dismal with a 10-13% 5 year survival. The best prognosis is associated with an early stage (Stage I and Stage II) however only 25% of patients present at these early stages. Over 50% of patients will have metastatic disease at presentation.

Etiology

Occupational exposures were the first recognized to be associated with the development of lung carcinoma. These include radon, arsenic, chromium among other work place chemicals. Despite this, it is believed that smoking is the largest single contributor to the induction of lung cancer. Smoking now accounts for 80-90% of all lung cancer cases, but only 20-30% of smokers will develop lung cancer. Several retrospective studies have found an association between the development of lung cancer and the quantity and duration of smoking. Most carcinogens tend to have a synergistic effect when combined with cigarette smoke. Passive or second hand smoke is considered to account for 25% of nonsmoker related lung carcinoma increasing the risk by 35-53%.

Pathophysiology

Bronchogenic cancers arise from basal or mucous cells on the surface epithelium of the bronchial tree. Other potential sites include neurosecretory cells such as Clara cells and Kultchitsky cells. The histologic subtypes of lung carcinoma can be divided into two main categories; small cell carcinoma (SCLC) and nonsmall cell carcinoma (NSCLC). NSCLC constitutes approximately 80% of lung cancers with the remainder being SCLC.

Current Concepts in General Surgery: A Resident Review, edited by William R. Wrightson.
©2006 Landes Bioscience.

Nonsmall Cell Lung Cancer

Nonsmall cell lung cancer represents the majority of lung neoplasms (80%).
- Adenocarcinoma (45%)
- Squamous cell carcinoma (33%)
- Large cell carcinoma (6%).

Adenocarcinomas tend to be more peripheral arising in the subsegmental bronchi away from the hilum while squamous cell carcinomas are centrally located.

Small Cell Carcinoma

SCLCs are typically widely disseminated at the time of diagnosis and therefore are rarely amenable to surgical resection. Only 10% of patients have disease that is determined resectable and in general are the subsets of patients with node negative disease. In contrast to NSCLC, SCLC is generally responsive to chemotherapy. Nearly 60% of patients will have a complete or partial response but will reoccur. The 5 year survival remains around 10%.

Other

These may contain one of several less comn forms of lung cancer. They include:
- Adenomas
- Carcinoids (neuroendocrine)
- Adenoid cystic carcinoma
- Mucoepidermoid carcinoma
- Mucus gland adenoma

Genetic Factors

Genetics plays a significant role in lung cancer. As with most cancer this is related to over or under expression of certain genes or loss of genotype. Lung cancers that are aneuploid have a 35% 5 year survival compared to a 61% survival if diploid. Loss of the 3p of chromosome 3 is also associated with a decreased survival and is found in 50-60% of NCSLC. Additional genetic factors are shown in Table 19.3. Many of these are potential targets for gene therapy with current Phase III trials ongoing.

Screening Issues

There has been significant interest in mass screening programs to detect lung cancer at an earlier stage. Use of sputum cytology and chest radiographs have fallen short and proven not to improve survival. Studies conducted in the 1970s at Memorial Sloan Kettering Lung Project and the Johns Hopkins Lung Project did identify more early stage lung cancers amenable to surgery; however, there was no statistically significant difference in mortality between screened and nonscreened groups. CT scanning has emerged as a potential means of identifying early lung cancers. Early studied indicate that it is more sensitive (27/1000 vs 10/1000) than sputum cytology and chest radiographs. There are currently several trails (National Cancer Institute) underway to evaluate CT scanning as a potential screening tool.

Clinical Presentation

Cough is probably the most common presenting symptom in lung cancer and is related to endobronchial erosion and irritation. Centrally located lesions may result in a change in a chronic cough, hemoptysis, pneumonia. More peripheral tumors may present with chest pain and or cough related to chest wall and pleural involvement.

19

Local extension of lung cancers results is varying presentations and syndromes. Invasion of the recurrent laryngeal nerve may result in hoarseness in up to 8% of cases. Dysphagia may be an indication of esophageal extension and is seen in 1-5% of presentations. Paraveterbral extention with involvement of the sympathetic nerve plexus results in Horner's syndrome (meiosis, ptosis, ipsilateral anhydrosis). Superior vena cava syndrome results from extrinsic compression of the superior vena cava. Patients present with jugular venous distention, edema of the face neck and arms.

Malignant pleural effusions are present in 30-60% of patients. If the effusion reoccurs after therapy, this represents a poor prognostic indicator.

Paraneoplastic syndromes occur in 10% of patients with lung cancer. Paraneoplastic syndromes associated with weight loss, neuromyopathies (Eaton Lambert syndrome) and elaboration of a variety of active peptides. Elevations in PTH-like peptide and calcitonin are associated with hypercalcemia, and pulmonary hypertrophy osteoarthropathy. Cushing's syndrome may present in as many as 38% of patients with bronchial carcinoid. Antidiuretic hormone (ADH) has been found elevated in as many as 50% of patients with only 5% showing symptoms of SIADH.

Diagnosis

Diagnostic Radiology

The most common reason for referral to a surgeon is the identification of a solitary pulmonary mass found on a chest radiograph. The most significant aid in the evaluation of a new pulmonary mass on chest radiograph is a previous chest radiograph for comparison. If the doubling time of the tumor is < 1 month it is likely infectious. If the doubling time is > 16 months it is likely to be benign. If the doubling time is between 1 and 16 months malignancy is more likely. Computed tomography (CT) is useful to determine the extent of tumor involvement and can detect enlarged lymph nodes. Lymph nodes > 1 cm have a 70% chance of being malignant. Without mediastinal node involvement, most patients are considered candidates for resection with a 5% false negative rate.

PET scan has emerged in recent years as a potential tool for distinguishing benign from malignant pulmonary nodules (Dewan). PET scanning has a sensitivity of 94% and a specificity of 80% in distinguishing benign from malignant pulmonary nodules.

Bronchoscopy

Bronchoscopy is useful to determine involvement in the bronchi and can provide diagnosis. If a tumor is visible, then washings and biopsy have a diagnostic yield of 94%. In contrast, if the tumor is not visible then the diagnostic yield falls to 66% (Sabiston).

Needle Biopsy

Transthoracic fine needle aspiration (TFNA) can be used to obtain a diagnosis; however a benign or indeterminate result is not exclusive owing to a high false negative rate. TFNA is generally reserved for poor operative candidates for thoracotomy who require a diagnosis for alternative treatment. There is a 30% rate of pneumothorax following biopsy.

19

Mediastinoscopy can be used to evaluate suspicious mediastinal nodes or provide a means of biopsy of centrally located lesions. For lesions in the aorticpulominary window, a Chamberlain procedure can be done to obtain a biopsy.

Thoracoscopy
Thoracoscopy is becoming more popular at instrumentation and optic technology had improved. VATS can provide diagnostic information as well as a means of therapeutic resection in certain cases.

Staging
See Tables 19.1 and 19.2.

Management
Unfortunately only 30% of patients are resectable at presentation with an increasing mortality rate. Surgery is the only curative treatment for lung cancer to date.

Preoperative Evaluation

Pulmonary Function Testing
The determination of resectability resides not only in technical feasibility but also on the determination that sufficient pulmonary reserve exists to tolerate a

---

Table 19.1.   Staging of lung cancer

Staging for Lung Cancer

Tumor (T)	
TX	Occult carcinoma (malignant cells in sputum or bronchial washings but tumor not visualized by imaging studies or bronchoscopy)
T1	Tumor 3 cm or less in greatest diameter, surrounded by lung or visceral pleura, but not proximal to a lobar bronchus
T2	Tumor > 3 cm in diameter, or with involvement of main bronchus at least 2 cm distal to carina, or with visceral pleural invasion, or with associated atelectasis or obstructive pneumonitis extending to the hilar region but not involving the entire lung
T3	Tumor invading chest wall, diaphragm, mediastinal pleura, or parietal pericardium; or tumor in main bronchus within 2 cm of, but not invading carina; or atelectasis of obstructive pneumonitis of the entire lung
T4	Tumor invading mediastinum, heart, great vessels, trachea, esophagus, vertebral body, or carina; or ipsilateral malignant pleural effusion
Nodes (N)	
N1	Metastases to ipsilateral peribronchial or hilar nodes
N2	Metastases to ipsilateral mediastinal or subcarinal nodes
N3	Metastases to contralateral mediastinal or hilar or to any scalene or supraclavicular nodes
Distant Metastases (M)	
M0	No distant metastases
M1	Distant metastases

19

Table 19.2. Staging groups for lung cancer

Stage	
Occult	TX N0 M0
Stage I	T1-2 N0 M0
Stage II	T1-2 N1 M0
Stage IIIa	T3 N0-1 M0 T1-3 N2 M0
Stage IIIb	T4 N0-2 M0 T1-4 N3 M0
Stage IV	Any T Any N M1

thoracotomy and resection. A patient with a preoperative FEV1 or FVC of > 30% usually can tolerate resection. A postoperative FEV1 of > 800 mL is also considered sufficient for resection. If a patient does not meet the criteria then a split lung perfusion study can be performed to determine the contribution of the involved segment of lung to be resected to the overall pulmonary function. Arterial blood gases should be done and have a PaO_2 > 55 and a $PaCO_2$ < 50. The diffusion capacity of carbon monoxide (DLCO) is also a useful measure of pulmonary function used preoperatively to predict morbidity.

Treatment
Treatment of Stage I and Stage II is primarily surgery.
1. Wedge Resection
2. Segmentectomy
3. Lobectomy
4. Pneumonectomy
5. Sleeve Resection

Treatment for Stage IIIA remains controversial; however induction therapy has had up to 30% of patients resectable. The remaining stages are managed primarily with radiation and chemotherapy.

Stage I
Stage I carcinoma of the lung can expect 3- and 5-year survival rates of approximately 85% and 70%, respectively. The most favorable group of patients with Stage I disease, those with T 1N 0 disease, experience 5-year survival rates of 80% to 85%. Neither chemotherapy nor radiation therapy is recommended after complete resection of Stage I lung cancer.

Stage II
Surgical therapy includes resection of the primary tumor with resection of the hilar, interlobar, lobar, and segmental lymph nodes. In addition, systematic mediastinal lymph node dissection is performed to exclude the presence of mediastinal metastases. Five-year survival rates of 40% to 50% are observed with a recurrence rate of greater than 50%. Postoperative radiation therapy may reduce the incidence of local and regional recurrence but does not affect overall survival.

Stage IIIa
This is a controversial area with current recommendations of resection combined with some form of adjuvant therapy.

19

Table 19.3.   Genetic factors in lung cacner

Oncogenes	Functions	Comment
K-ras	Signal transduction	90% of lung cancers
HER-2	Chemoresistance	Decreased survival
BLC-2	Proto-oncogene inhibits apoptosis	Increased survival when present
MYC	DNA binding and cell cycle regulation	
Tumor suppressor		
p53	Cell cycle control and proliferation	
RB	Cell cycle control and proliferation	Decreased survival

Stage IIIb and Stage IV

Patients with Stage IIIb NSCLC (T 4 or N 3 disease) are generally considered unresectable. Stage IV have metastatic disease and are considered unresectable with chemotherapy and radiation therapy the only available treatment (Tables 19.4 and 19.5).

Questions

1. Paraneoplastic syndromes are associated with ALL of the following except:
   a. Cushings syndrome is more often associated with bronchial carcinoids than squamous cell carcinoma.
   b. Less than 5% of patients with small call carcinoma will develop Cushing's syndrome.
   c. Hypercalcemia is more common in small cell cancer than squamous cell cancer.
   d. Hyponatremia secondary to antidiuretic hormone is found in 5% of patients with lung cancer.
   e. Hypercalcemia is usually caused by a PTH like peptide secretion.
2. All of the following regarding CT biopsy and thoracoscopic biopsy of a solitary pulmonary nodule (SPN) are true except:
   a. Postoperative complication rates are equal.
   b. CT biopsy is the definitive management of a SPN.

Table 19.4.   Neoadjuvant trials for nonsmall cell lung cancer using induction therapy followed by surgery

Study	Stage	Chemotherapy	Radiotherapy	Median Survival (Months)	2 Year Survival
Lung Cancer Study Group (1991)	IIIA, IIIB	Cisplatin, 5-Fluorouracil	3000 cGy (15 fractions)	11	8
Southwest Oncology Group (1993)	IIIA, IIIB	Cisplatinin and Vinblastine	4500 cGy (15 fractions)	IIIA - 13, IIIB - 17	IIIA - 37, IIIB - 39

19

Table 19.5. Survival by stage

Stage	5 Year Survival
Stage I	67
Stage II	44
Stage IIIA	22
Stage IIIB	5
Stage IV	5

   c. False positive rates are equal between each procedure.

   d. Thoracoscopic biopsy can be performed safely with a FEV1 of 1.5L.

   e. The number of indeterminate diagnoses is higher with CT than thoracoscopy.

3. A 55-year-old man with a long smoking history has a 2 cm irregular noncalcified lesion in the right upper lobe detected on chest radiograph. CT identifies the same peripheral nodule. He is asymptomatic, except for some fevers at night. Management should include:

   a. Careful observation with serial x-rays

   b. Skin tests for TB and fungi

   c. Excision of the lesion

   d. Trial of antibiotic therapy

   e. Mediastinoscopy

References

1. Surgery of the Chest Sabiston DC and Spencer FC. 6th ed. Philadelphia: W.B. Saunders Company, 1996.

2. Gould MK, Maclean CC, Kuschener WG et al. Accuracy of positron emission tomography for the diagnosis of pulmonary nodules and mass lesions: A meta-analysis. JAMA 2001; 285:914-924.

3. Patz EF, Goodman PC, Bepler G. Current concepts: Screening for lung cancer. N Engl J Med 2000; 343:1627-1633.

4. Surgical education and self assesment program. 10th ed. American College of Surgeons, 1999-2001.

Minimally Invasive Coronary Artery Bypass Grafting

History

Coronary artery bypass grafting has withstood the test of time, aspirin, HMG coreductase inhibitors and angioplasty. Three major randomized studies, the Coronary Artery Surgery Study (CASS), the Veterans Administration Cooperative Study Group (VA) and the European Coronary Surgery Study (ECSS), and four smaller randomized trials conducted between 1972 and 1984 provide the most reliable outcome data comparing medical and surgical therapy. A recent meta-analysis of seven randomized trials show that surgically treated patients at high risk (4.8 percent annual mortality) and moderate risk (2.5 percent annual mortality) experience clinically and statistically enhanced survival at 5, 7, and 10 years.

Indications for Coronary Artery Bypass Grafting (CABG)

Patients with left main equivalents (proximal left anterior descending and proximal circumflex artery stenosis), three-vessel disease (irrespective of left ventricular dysfunction), and double-vessel disease with left ventricular dysfunction all appear to have a survival benefit from surgical revascularization in comparison with both medical therapy and angioplasty.

For patients with less-extensive atherosclerosis and preserved left ventricular function, no distinct advantage in terms of survival has been demonstrated for one therapy (medical, angioplasty, or surgery) over another. A high proportion of lower-risk patients included in revascularization trials are alive and angina-free regardless of the initial mode of revascularization for up to 5 years after initial therapy. However, patients who initially undergo CABG surgery have fewer subsequent revascularization procedures, antianginal medicines, or cardiac rehospitalizations than those on medical therapy or those treated initially by angioplasty. A recent study showed only 8% of CABG treated patients required revascularization compared to 34% of those receiving PCTA.

Among patients with unstable angina, electrocardiographic changes with pain, and evidence of left ventricular dysfunction are all indicative of severe ischemia and/or left ventricular dysfunction and strongly warrant consideration for revascularization depending on coronary artery anatomy. Patients with postinfarction angina are a high-risk group with severe ischemia and left ventricular dysfunction for whom the likelihood that revascularization will be beneficial is strong and logical.

Complications of CABG and CPB

For 1997, in a dataset comprising 161,018 primary CABGs, the overall operative mortality was 2.8%. Major complications included Q-wave myocardial infarction (1.1%), adverse neurological events (5.4%), pulmonary complications (11.2%), acute renal failure (3.1%), and sternal wound infection (1.4%). Stroke was observed in 3.1% of patients and cognitive dysfunction or seizures were observed in 3.0%.

Cardiopulmonary bypass (CPB) carries with it inherent risks. In addition to the risk of microemboli, CPB induces a total body inflammatory response caused by the activation of the complement system due to contact of the blood with the artificial surface of the CPB circuit. All organs are affected to a varying degree, potentially leading to dysfunction and/or damage of the brain, lungs, heart itself, bowel, kidneys, and coagulation system. Nonpulsatile flow is one of the mechanisms which, in combination with the inflammatory response and the release of free radicals, is thought to be responsible for postoperative renal failure.

Off-Pump Coronary Artery Bypass (OPCAB)

Off-pump surgery is a technique that allows the surgeon to perform a bypass procedure without the use of CPB. Consequently, patients do not experience the global inflammatory response caused by CPB, which disrupts the body's physiologic balance. OPCAB procedure utilizes a traditional sternotomy. The greatest advance in OPCAB came with the introduction of regional mechanical stabilisers such as the CardioThoracic Systems Ultima device and the Utrecht Octopus in the mid 1990s. These devices consistently reduced the motion of the target area facilitating the anastomosis.

Table 19.6.   Indications for OPCAB

1. Isolated LAD, RCA stenosis or both not amenable to PCTA
2. High risk patients (advanced age, severe left ventricular dysfunction, prior stroke, and severe pulmonary and renal dysfunction)
3. Reoperations

Indications
(Tables 19.6, 19.7)

Current Problems

Generally OPCAB is well tolerated; however, it may occasionally provoke arrhythmia and hemodynamic instability. The interruption of the flow of the right coronary artery is known to provoke some of these complications. The use of an intracoronary shunt when performing the anastomosis may attenuate this response. Most shunts are silicone tubes, 10 to 30 mm in length by 1.0 to 3.0 mm in diameter, designed to be inserted through the arteriotomy into the coronary lumen. The clinical value of shunts is questioned since they are cumbersome to use and, with respect to the shunt, blood flow through the shunt is only 30-50% of the native coronary flow.

Hemodynamic instability may be encountered with manipulation of the beating heart. The left anterior and diagonal arteries are relatively easy to approach. This is not true of the right coronary and circumflex arteries. Posterior grafting may require significant displacement of the heart but seems to be relatively well tolerated in most patients. To mitigate instability, the patient is placed in steep Trendelenburg position and rotated to the right.

OPCAB has significant advantages in patients for whom cardiopulmonary bypass and aortic manipulation presents an increased risk (renal, pulmonary, hepatic, and neurologic injury and/or insufficiency, aortic atheroma/calcification) and multivessel revascularization is required. The disadvantage of this approach is the technical difficulty of bypassing the posterior and lateral wall vessels and the need for a full midline sternotomy.

MIDCAB Grafting

The most common features of the MIDCAB procedure are avoidance of full sternotomy, CPB, and aortic manipulation. The premise for adopting this procedure is reduction of morbidity, costs, and length of hospital stay without compromising the quality of the surgical procedure.

Table 19.7.   Contraindications for OPCAB

Relative Contraindications

1. Vessel diameter < 1.5 mm
2. Severe vessel calcification
3. Intramyocardial vessel
4. Severe cardiomegaly

19

Anterior MIDCAB is performed through a small minithoracotomy in the fourth intercostal space underneath the nipple. Grafting of mid-left anterior descending (LAD) and diagonal branches can be performed by this approach. The anterolateral MIDCAB approach is used to bypass the ramus intermedius and first obtuse marginal branch of the circumflex system. An incision is made in the third interspace from the midclavicular to anterior axillary fold. Bilateral internal mammary artery (IMA) grafting to the LAD and OM has been made possible by thoracoscopic mobilization of right IMA. Lateral MIDCAB is performed with a small lateral thoracotomy through the fifth or sixth intercostal space. Marginal branches of the circumflex artery are accessed through this incision. Right bucket handle partial sternotomy is usually performed through a lower right partial sternotomy for right IMA to mid-right coronary artery (RCA) anastomosis. Sometimes electrocardiographic changes and hemodynamic instability are the major problems, especially in case of proximal RCA stenosis.

Extrathoracic MIDCAB

The distal RCA and posterior descending artery are approached via a 2.5-inch subxiphoid incision. The xiphisternum is excised, and bilateral release of the costal diaphragmatic attachments helps provide better exposure of the target vessels. The right gastroepiploic artery is a frequently used conduit for this procedure.

The greatest technical limitation for MIDCAB techniques is multivessel disease. To bypass multiple coronary arteries using multiple arterial conduits, the transabdominal approach is performed through a 3-inch chevron epigastric incision. Division of bilateral recti with release of bilateral diaphragmatic attachments widely opens the substernal window. Bilateral mammary arteries and the right gastroepiploic artery are available conduits with this incision. The LAD is approached above the diaphragm, while the distal RCA, posterior descending artery, and posterial lateral marginal artery are approached through the diaphragm after mobilizing the left lobe of the liver.

Results

OPCAB has grown in application from a small number in 1995 to 10% in 1999, and is expected to be 50% of all CABG by 2005. Nonrandomized studies have shown mortality 0-3.8%, stroke 0-2%, renal failure 0-6.7%, atrial fibrillation 8-30%. Several studies have shown a postoperative angiographic graft patency was 99%, with perfect graft patency (< 50% stenosis in the proximal or distal anastomosis or graft body) present in 91% to 98.5%.

References

1. deJaegere PP, Suyker WJL. Off-pump coronary artery bypass surgery. Heart 2002; 88(3):313-318.
2. Takaro T, Hultgren HN, Lipton MJ et al. The VA cooperative randomized study of surgery for coronary arterial occlusive disease, II: Subgroup with significant left main lesions. Circulation 1976; 54 suppl:III107-III117.
3. The veterans administration coronary artery bypass surgery cooperative study group. Eleven-year survival In the veterans administration randomized trial of coronary artery bypass surgery for stable angina. N Engl J Med 1984; 311:1333-1339.
4. Duhaylongsod, Francis G. Minimally invasive cardiac surgery defined. Arch Surg 2000; 135(3):296-301.

CHAPTER 20

Endoscopy

C. Adam Conn and William R. Wrightson

Endoscopy

Background

The advent of the fiberoptic and subsequently endoscopic technology has allowed the surgeon to examine the upper and lower GI tract. This has provided a cost-effective means of cancer surveillance, monitoring of gastrointestinal pathology as well as enhancing minimally invasive therapeutic options. Endoscopy is considered both a diagnostic and therapeutic intervention for this reason. The surgeon should have a basic understanding of the scope mechanism as well as the most common disease processes that the scope can be used to diagnose and treat.

Upper Endoscopy

Technique

Upper endoscopy is used for the evaluation of the esophagus, stomach, duodenal bulb and first portion of the duodenum. This is performed with a flexible scope with conscious sedation. It allows you to exclude other disease, document the presence of esophageal injury, and to score the degree of injury using one of the scoring systems available. In Barrett's esophagus, the most advanced form of this disease, biopsies should be performed to confirm metaplasia and exclude dysplasia. This can be done using an instrumentation port and biopsy forceps.

Therapeutic Potential

In the case of an **acute upper GI bleed**, endoscopy can be both diagnostic and therapeutic allowing the operator to advance to the initial steps of definitive therapy. However, endoscopy in this acute setting carries a three-fold increase in risk. When compared to elective endoscopy. More notable risks include arterial desaturation and in the hypotensive patient, aspiration.

Peptic Ulcer

Bleeding from a **peptic ulcer** remains the most common source of upper GI bleed accounting for greater than 50%. This can be controlled approximately 94% of the time using endoscopic techniques. Visualization of an ulcer itself can be an important predictor of rebleed. Ulcers have four recognized appearances: clean base; a flat, pigmented spot, which at times may be purple; an adherent clot; a visible vessel; or active bleeding. These are all considered to have an increased risk of hemorrhage. A clean ulcer base rarely bleeds; a flat pigmented base will bleed in about 10% of patients; an adherant nonbleeding clot rebleeds in about 20% of patients; and a visible vessel carries about a 40-80% rebleed rate.

Acute Esophageal Varices

In acute **esophageal varices** sclerotherapy and banding have become the initial prefered treatment. Sodium morrhuate and sodium tetradecyl sulfate are used as sclerosing agents. Each varix is usually injected just above the esophagogastric junction and 5 cm proximal to it. Additional treatments should follow in 5 to 6 days. The success rates of such a procedure in emergent situations has been as high as 85%.[1] However gastric varices have not shown such promise.[5] Sclerotherapy should be considered a failed intervention when two sessions have been completed and hemorrhage has still not subsided. In such cases, mortality rates exceed 60% unless urgent surgery is undergone. Side effects of sclerotherapy include pain, fever, ulceration, esophageal perforation, worsening of bleeding varices, and aspiration pneumonia. In combination these can account for a 1-3% mortality. Some investigators have shown banding therapy to be as effective as sclerosis and have fewer complications.[6] It requires fewer sessions and has a lower incidence of rebleed and mortality.

Chronic Esophageal Varices

In chronic esophageal varix therapy endoscopy is in some cases used with the goal to eradicate varices all together. This has been shown successful in about two-thirds of patients. These patients should, however, be followed closely and rescoped at 6 month to 1 year intervals to evaluate for the likelihood of rebleed. Recurrent bleeding occurs in about half of these patients with the highest rate in the first year and decreases by about 15% per year after that. One-third of these patients will fail secondary to rebleed. Endoscopic therapy should be abandoned with uncontrolled hemorrhage, multiple major episodes of recurrent rebleed of gastric varices and PHG.

Stress Gastritis

Bleeding from **stress gastritis** can be endoscopically diagnosed and controlled in much the same manner. The use of heater probes, injection therapy, and laser or electrocoagulation have all proven useful for therapy under the direct visualization of the scope.

Gastric Cancer

Flexible fiberoptic endoscopy is the gold standard of the diagnosis of **gastric cancer**. It can appear as a polypoid lesion, plaque like lesion or a shalow ulcer in the mucosa. Each time one of these is seen in the stomach a minimum of seven biopsies should be obtained for a greater than 98% diagnostic accuracy. Brushing can also be used in conjugation with biopsy for a diagnostic accuracy of almost 100%, or brushing can be used alone. Both of these should be done at the edge of an ulcer, not the crater base.[1]

Endoscopic Retrograde Pancreatography (ERCP)

In acute pancreatitis, ERCP is used in recurrent episodes without an identifiable cause. It has identified the cause in about 50% of such patients.[1] ERCP is usually performed after the symptoms have resolved. Some expected findings include pancreas divisum, stenosis of the ampula of Vater, and in some cases focal pancreatic ductal stenosis.[7] There is no role for ERCP in the initial onset of acute pancreatitis.

In chronic pancreatitis ERCP can demonstrate ductal abnormalities such as dilations, strictures, and pseudocysts which cannot otherwise be visualized. The classic chain of lakes appearance, with alternating dilation and stricture is the most classic

20

finding of chronic pancreatitis but is less common that ductal dilation alone. Chronic pancreatitis can be safely excluded with a normal ERCP.[7]

ERCP also provides the opportunity to intervene with stenting, sphincterotomy, and stone extraction procedures to allow for drainage of the pancreas. This can provide dramatic pain relief to the patient with acute pancreatitis. However, long term data on pain relief for patients with chronic pancreatitis is not yet available. Stents, however, are used as temporizing measures with possible surgical intervention still in sight. Complications from these procedures are exacerbation of pancreatitis, stent migration, stent breakage, stent occlusion, and pancreatic ductal changes.

Rigid Proctoscopy

This method can be diagnostic for colon and rectal pathology; however it has been replaced for routine cancer surveillance by endoscopy. This is still used by many surgeons for perioperative management of those patients with rectal pathology. It precisely establishes the distance between a rectal tumor and the anus. Used during surgery it can define the rectal stump that has retracted after a Hartman pouch procedure. Postoperatively this can be used to view and dilate a low coloproctostomy. It can be used to study pelvic outlet obstruction and demonstrate nonrelaxing pubis rectalis or rectal intussusception.

Flexible Sigmoidoscopy

Flexible sigmoidoscopy is preferred over colonoscopy because it can be done without sedation. It is preferred over rigid sigmoidoscopy because it provides a working channel. Indications include evaluation of some forms of colitis that affect the distal colon, such as ischemic, actinic, granulomatous, and ulcerative. Some still recommend this for cancer surveillance, although this has in most areas been replaced by the flexible colonoscope secondary to its ability to survey the entire colon in one visit and discover more proximal lesions.

Colonoscopy

This technique uses a video camera mounted in the end of the scope to render recordable images and color photos of the colonic mucosa. It requires that the colon be devoid of stool and the patient be sedated. The main advantage of this technique over barium enema (BE) is that is allows for surgical intervention. Tumors can be biopsied, cauterized and even removed using different instruments which can be inserted into the endoscope port. However it requires more technical skill to perform than BE. It may even prove to be impossible to perform in some patients, women (31%) usually more often than men (16%), due to the inherently larger colons of some women. Diverticular disease can also complicate the exam. Sensitivity of colonoscopy is not 100%, usually due to rapid transit time across certain areas of the colon, such as the flexures and the rectosigmoid junction.

Other Uses

Endoscopes are also being used to evaluate and treat pneumonia. Bronchoscopes can be used to remove excess secretions and mucus plugs in a therapeutic manner. It can also be used to remove foreign bodies from the airway.

The endoscope can be used to evaluate and remove bladder tumors, sinus tumors, nasal tumors, aid in the placement of esophagel stents in esophageal carcinoma, and place percutaneous gastrostomy tubes for feeding. However that is beyond the scope of this chapter and could encompass an entire book.

References

1. Townsend: Sabiston textbook of surgery. 16th ed. © 2001 W.B. Saunders Company, 216, 292-294, 334-335, 646-659, 757-758, 816-834, 843-854, 929-969, 1065-1073, 1080-1082, 1089-1092, 1112-1142, 1435-1443, 1492, 1493, 1485-1486, 1678.

2. Williams RA, Vartany A, Davis IP et al. Impact of endoscopic therapy on outcome of operation from bleeding peptic ulcers. Am J Surg 1993; 166:712-715.

3. Benders JS, Bouwman DL, Weaver DW. Bleeding gastroduodenal ulcers-improved outcome from a unified surgical approach. Am Surg 1994; 60:313-315.

4. Brugge, William R, Van Dam. Jacques: Pancreatic and biliary endoscopy. N Engl J Med 1999; 341(24):1808-1816.

5. D'Amico G, Pagliaro L, Bosch J. The treatment of portal hypertension. A metanallysis review. Hepatology 1995; 22:332.

6. Stiegmann GV, Goff JS, Michaletz-Onody PA et al. Endoscopic sclerotherapy as compared with endoscopic ligation for bleeding esophageal varices. N Engl J Med Jun 4 1992; 326(23):1527-32.

7. Vitale Gary C MD. Advanced interventional endoscopy. Am J Surg 1997; 173(1):21-25.

8. Cameron John L. Current surgical therapy. 7th ed. Copyright 200, Mosby Inc., 1 347, 462, 463, 496, 497, 516-521, 529, 530, 543, 545, 559, 560, 1083.

9. Saunders BP, Fukumoto M, Halligan S et al. Why is colonoscopy more difficult in women? Gastrointest Endoscosc 1996; 43(2 pt 1):124-126.

CHAPTER 21

Orthopaedic Trauma

James C. Dodds and S. Matthew Rose

Orthopaedic Trauma

Introduction

On average, 170,000 people in the United States are injured per day, 400 of whom die as a result of these injuries. Accidental injury is the greatest cause of death in people aged 1-44.[1] Extremity injuries are found in 40-60% of all traumatic brain injured patients.[2] Having an organized, multidisciplinary approach to polytraumatized patients greatly decreases the associated mortality.[3]

Treatment of the multiply injured patient is a challenge and requires a multidisciplinary approach with adequate resources and effective resuscitative measures. Success in treating these patients begins at the accident scene with identification of injuries, temporary stabilization, and judicious transportation to trauma centers. In the hospital setting, assessment, resuscitation, and definitive treatment followed by rehabilitation and return to function must be performed in an organized, systematic fashion.

In the last 30 years there has been extensive literature supporting early stabilization and fixation of pelvic and long bone fractures in the polytraumatized patient. This body of literature reports benefits of early stabilization such as decreasing morbidity and mortality, decreasing the incidence of pneumonia and ARDS, increasing mobilization, and decreasing the number of ICU, as well as, overall hospital days.[1]

Evaluation

The ABCs of assessment and resuscitation per the ATLS protocols always come first. Standard to all trauma evaluations are the cross-table C-spine, AP chest, and AP pelvis x-rays. As the patient is examined, any gross deformity should be noted and suspected long bone injuries should be x-rayed with two views 90° angulated from each other. To insure associated injuries are not missed, the joint above and below must be included in the x-ray evaluation of a long bone fracture. Any soft tissue wound should be evaluated for possible open fracture or joint arthrotomy. Further evaluation modalities can include appropriate CT scans, MRI, or ultrasound studies. Repeated exams throughout evaluation and treatment are very important in any trauma situation to ensure that no injuries are missed. With the constellation of factors in the polytraumatized patient, 5-20% of patients have unrecognized injuries after the initial evaluation.

There are many fracture classification systems in use to describe closed long bone fractures. The most important issue is to be able to accurately describe the bones involved and their related fractures. Fractures are described naming the involved bones and then describing the fracture site in terms of angulation, displacement, rotation, comminution, and finally whether it is open or closed. Displacement is always used to describe the distal fragment in relation to the proximal fragment.

Current Concepts in General Surgery: A Resident Review, edited by William R. Wrightson. ©2006 Landes Bioscience.

Orthopaedic emergencies that must be identified and addressed include pelvic injuries, open fractures, compartment syndrome, and joint dislocations.

Pelvic Trauma

Pelvic fractures are a major concern to the trauma team because they are often associated with other injuries including traumatic brain injury, abdominal injuries, thoracic injuries, and associated musculoskeletal injuries. High energy pelvic fractures have an associated 15-25% mortality rate, have an increased incidence of aortic rupture of eight times normal, and 60-80% have associated musculoskeletal injuries.[1] Associated injuries directly related to pelvic fractures include:

- Hemorrhage: Can occur in up to 75% of high-energy pelvic fractures. Bleeding is most commonly from the pelvic venous plexus but can also be due to named arterial injury (internal, external iliac arteries and their branches).[1] The retroperitoneal space can contain up to 4 L of blood from these sources.
- Visceral rupture: due to bony spicules or direct pressure from injury rupturing urethra, bladder, uterus, vagina, or rectum. Urogenital injury is seen in 12% of high-energy pelvic fractures.[1] Rectal and vaginal exams must be performed to rule out any occult open fracture due to bony spicules associated with pelvic fractures.
- Neural disruption: the lumbosacral trunk and the L5 nerve root are especially vulnerable with sacral ala fractures and sacroiliac joint dislocations. The lumbosacral trunk is injured in 8% of high-energy pelvic fractures.[1]

Evaluation

Signs of pelvic fracture on physical exam include ecchymosis in the perineal or lower abdomen, unequal anterior superior iliac spines, unstable movement of the pelvis with manual traction to one leg, and unstable pelvic tilt and compression tests. It is important that tests to check the stability of the pelvis only be performed by one person secondary to the possibility of causing further injury with movement of the fragments.

Ninety percent of all pelvic ring injuries can be detected on an AP pelvis x-ray.[1] When pelvic ring fractures are suspected, then additional views should include a pelvic inlet view (45° caudal tilt) and a pelvic outlet view (45° cranial tilt). A CT scan of the pelvis can also be helpful in evaluating comminution and sacral involvement. Once a bony pelvic injury is identified, the primary goal becomes stabilization of the pelvic ring to prevent further injury, decrease hemorrhage, and increase mobilization of the patient.

Classification

There are several classification schemes for pelvic ring injuries, but the Young and Burgess system[4] based on mechanism of injury seems to be the most simplistic. Helps to identify risk of hemorrhage and dictate resuscitation in polytraumatized patients, and has four major categories:

- Lateral compression (LC): typically seen in the classic "rollover" motor vehicle accident. Usually composed of oblique anterior ring fractures with an anterior sacral impaction fracture or iliac wing "crescent fractures" posteriorly.[5] Usually see internal rotation of one hemipelvis. High incidence of associated traumatic brain injuries and abdominal injuries.

- Anteroposterior compression (APC): usually composed of vertical ramus fractures or pubic symphysis widening ("open book" pelvis). As the injury worsens there is progressive loss of anterior then posterior sacral ligaments with resultant neural and/or vascular damage. Can be associated with very high blood loss. There is an increase incidence of CNS, abdominal, and visceral injury.
- Vertical shear (VS): vertical displacement of one hemipelvis; associated with increased blood loss.
- Combined mechanisms: combined patterns of each of the three types of pelvic ring fractures.

Treatment

Hemorrhage resulting from pelvic fractures needs to be addressed rapidly in order to facilitate stabilization of the patient. Emergent measures consist of MAST trousers, pelvic binders, or a tightly wrapped sheet around the pelvis during transportation. After full evaluation of the patient, options include: exploration with ligation, open reduction internal fixation of fragments, external fixation, and angiography with embolization of bleeding vessels. Angiography is an excellent modality where resources of the institution allow. External fixation is a good option for anterior stability of the pelvic ring and can decrease further injury, cancellous bleeding, and facilitate mobilization of the patient postoperatively. Fixation of the posterior ring, such as percutaneous SI joint screws, is often required to augment anterior fixation.

Open Fractures

Introduction

Historically, limb-and-life threatening injuries were only generated by warfare and natural catastrophes, but the industrial and technological advances of the past 200 years has exposed the human body to forces that the cannot be withstood. Approximately 60% of open fractures are associated with other major injuries to central nervous system, cardiothoracic and abdominal injuries, or other fractures or musculoskeletal injuries.[1]

Open fractures pose difficult problems in polytraumatized patients secondary to their emergent need to adequately debride them to prevent infection and possible sepsis. Severe open fractures can certainly be limb- and possibly even life-threatening injuries. The entire trauma team must then communicate effectively regarding needs for surgical intervention and the risks for general anesthesia pending patient status.

Classification

Again, several classification systems are available but the classic system that is simple and aids in treatment decisions is the Gustilo classification that reports three major types and considers mechanism, degree of soft tissue involvement, fracture pattern, and level of contamination.[6]

- Type I: Wound is less than 1 cm long. No sign of crush injury. Usually a clean puncture type wound and signifies an "inside/out" injury where a spike of bone punctures the skin. Little comminution of the fracture is present.
- Type II: Wound is larger than 1 cm, but no extensive soft tissue damage or flap exists. May be mild contamination, comminution, and may show signs of mild crush injury.

- Type III: There is extensive soft tissue damage with a high degree of contamination. Usually a result of high-energy trauma resulting in significant comminution and instability. Type III fractures are further classified into three subgroups.
    - III-A: There remains adequate soft tissue covering of the bone regardless of soft tissue damage.
    - III-B: Extensive soft tissue injury with massive contamination, muscle loss, fracture comminution, and periosteal stripping. After debridement a local or free flap is required to cover resultant exposed bone.
    - III-C: Involves any open fracture associated with arterial injury that must be repaired regardless of soft tissue injury.

Treatment

The importance of early identification of open fractures is paramount to treatment options, prevention of infection, and overall outcome. Incidence of infection is directly proportional to the degree of soft tissue injury from 0-7% in type I open fractures to 25-50% in type IIIC open fractures.[6] Studies have shown that cultures taken in the emergency department are positive in 70% of open fractures. Therefore the initiation of intravenous antibiotic therapy as early as possible is of great importance.[1] Current literature supports the use of a first-generation cephalosporin for 2-3 days for Gustilo type I and II injuries. For type III open fractures an aminoglycoside should be added for 3-5 days. And in farm-related injuries, vascular compromise, or injuries related to severe crush, penicillin may be added to cover *Clostridium perfringens* and *C. septicum*.[1,3,6]

Open fractures require immediate extensive and meticulous debridement with copious irrigation performed in the operating room. Antibiotic bead chains or a bead pouch technique can be beneficial to help fight infection. Stabilization of the fracture fragments can be temporary or definitive, and the need for repeated debridements or flap coverage may affect which fixation modality is used. Fixation options include splinting and casting, traction, open reduction and internal fixation with plates and screws, intramedullary devices, and external fixation.

Compartment Syndrome

Compartment syndrome is a condition typified by raised pressure within a closed space with the potential to cause irreversible damage to the contents of the compartment.[1] This condition can be present in any portion of the body in which there is a closed compartment: most commonly in the leg but can also be seen in the thigh, forearm, hands, and feet. Causes range from constrictive dressings, pneumatic antishock garments, fractures, tight casts, to vascular injuries. Crush injuries can also cause compartment syndrome; however, many now consider severe crush injuries to be an altogether different entity.[1] Do not overlook the possibility of compartment syndrome in open fractures as it occurs in 6-9% of all open fractures.[1]

Evaluation

There are frequently delays in diagnosis of compartment syndrome secondary to head injuries, level of consciousness, and other musculoskeletal injuries that may mask the symptoms. Therefore, there must be a high index of suspicion for compartment syndrome in evaluating polytrauma patients.

On exam, the compartment will be tense to palpation. The skin is often tight with a shiny appearance. The hallmark finding in compartment syndrome is pain disproportionate to that expected for the injury or to exam. Pain with passive stretch of the muscles in the compartment in question is a reliable indicator.[1] Late findings include paralysis (muscle weakness), pallor, and paresthesia. Pulses are often strong even in florid, acute compartment syndrome and therefore are not a reliable sign to help guide treatment options.[1]

Compartment pressures can be measured in a variety of ways and several different thresholds for fasciotomy from direct compartment pressure to a difference between diastolic and compartment pressure have been described. A commonly accepted threshold is compartment pressure of 30-35 mm Hg.

Treatment

In patients that exhibit signs and symptoms of compartment syndrome, the patient should be taken directly to the operating room for fasciotomy and compartment pressures are not needed. The standard treatment consists of releasing the compartments in question along the full length of the compartment. Fasciotomies are usually left open 3-5 days and can be closed using delayed primary closure, secondary closure, or skin grafts.

Dislocations

Relocation of a joint is very important to decrease the incidence of long-term disability and morbidity. Even with immediate attention and reduction, there may be damage of articular cartilage, joint capsule, ligaments, or neurovascular structures. These factors lead to increased incidence of ectopic ossification, post-traumatic arthritis, and avascular necrosis. The amount of time from injury until relocation directly correlates with outcome. This point is exemplified in the hip where the incidence of avascular necrosis is 17.6% when reduced within 12 hours compared with 56.9% when reduced after 12 hours.[3]

In the trauma setting, dislocations are most frequently seen in the hip but can also involve shoulder, elbow, wrist, knee, patella, and ankle. Acute dislocations of the knee are especially correlated with vascular injury and an arteriogram should be considered after relocation even in the presence of good pulses.

Evaluation

At least two x-ray views of the joint in question must be obtained to determine the exact direction of dislocation, which dictates reduction options. The joint surfaces and peri-articular segments of the bones involved must be evaluated for fracture. Neurovascular status should be evaluated distal to the dislocated joint and documented prior to relocation maneuvers.

Treatment

Most dislocated joints can be treated in a closed fashion with intravenous analgesia and sedation. Occasionally, patients must be taken to the operating room for general anesthesia to aid in reduction and may have to have open reduction if that fails. Care must be taken in relocation, as certain maneuvers are associated with high risks of fracture.

Once relocated, range of motion in the joint should be tested to assess for stability and possible ligamentous damage. The neurovascular status distal to the joint should be reevaluated. CT scan is frequently obtained in cases of fracture dislocation to assess articular fragments and joint congruity. MRI is beneficial to assess surrounding soft tissues and structural supports. Finally, an arteriogram may be required to assess vascular status.

Selected Reading

1. Browner BD, Jupiter JB, Levine AM et al. Skeletal trauma: Fractures, dislocations, ligamentous injuries. 2nd ed. Philadelphia: W.B. Saunders Company, 1998:1.
2. Kushwaha VP, Garland DG. Extremity fractures in the patient with a traumatic brain injury. J Am Acad Orthop Surg 1998; 6:298-307.
3. Canale TS. Campbell's operative orthopaedics. 9th ed. St. Louis: Mosby, 1998:3.
4. Burgess AR, Eastridge BJ, Young JWR. Pelvic ring disruptions: Effective classification system and treatment protocols. J Trauma 1990; 30:848-856.
5. Turen CH, Dube MA, LeCroy CM. Approach to the polytraumatized patient with musculoskeletal injuries. J Am Acad Orthop Surg 1999; 7:154-165.
6. Gustilo RB, Merkow RL, Templeman D. The management of open fractures. JBJS 1990; 72-A(2):299-304.

CHAPTER 22

Head and Neck Surgery

Steven L. Goudy

Cancer of the Larynx

Background
This is the second most common site of cancer, behind oral cavity. It is one of the most curable cancers of the upper aerodigestive tract.

Epidemiology
These cancers are more common in males (4:1 males), 60-70 years old. The majority occur in the supraglottis and glottis, subglottis uncommon with tobacco and alcohol are causative factors.

Pathology
The majority (95%) are squamous cell carcinoma.

Anatomy
Supraglottis—contains epiglottis, preepiglottic space, aryepiglottic folds, arytenoids, false cords, and half of the laryngeal ventricle (cavity between false and true cords). Has a good lymphatic supply

Glottis—contains lower half of laryngeal ventricle, true vocal cords, and extends 1 cm below the ventricle; has a poor lymphatic supply

Subglottis—area 1 cm below ventricle to inferior aspect of cricoid; has good lymphatic supply

Paraglottic space—area of cancer spread between supraglottic and glottic surface and the thyroid ala, pyriform sinus, and quadrangular membrane of the aryepiglottic folds

Treatment
Treatment is with chemotherapy and radiation therapy and has been shown to be as effective as surgery in treating lower stage cancers of the glottis with vocal preservation in at least 50%. Complications include xerostomia, dental carries, salivary fistula, carotid erosion. Surgical resection is an option but patients must have a good pulmonary reserve and must address both necks with surgery or radiation therapy.

Prognosis
Larynx 5 year survival—stage I 85-95%; stage II 65-75%; stages III and IV 40-60%

Supraglottis 5 year survival—stage I 90-95%; stage II 80-90%; stages III and IV 20-40

Current Concepts in General Surgery: A Resident Review, edited by William R. Wrightson. ©2006 Landes Bioscience.

Cancer of the Nasopharynx

Background
The most common presenting symptoms are unilateral cervical lymphadenopathy, nasal obstruction, hearing changes (conductive from blockage of eustachian tube).

Epidemiology
These are rare and occur 1:100,000 with average age of 50 years old. They are mose common in males (3:1 males), increased incidence in Southeast Asia may be related to salting of fish, Epstein Barr virus (EBV) strongly associated—check IgA titers to viral capsid antigen and early antigen.

Histopathology
- Type I keratinizing squamous cell carcinoma—no EBV titer, more resistant to treatment
- Type II nonkeratinzing squamous cell carcinoma—anti-EBV titer
- Type III undifferentiated carcinoma—anti-EBV titer

Management
The primary modality is XRT with 66-70 gray to primary site and 60 gray to the neck

Prognosis
- Stages I + II 5 year survival 50-90%
- Stages III + IV 5 year survival 17-60%
- Recurrences occur in 2-3 years in 90%

Cancer of the Oral Cavity

Background
The oral cavity includes vermilion of lips to hard/soft palate junction and circumvallate papillae.

Epidemiology
It represents 6% of all cancers, 30% of all head and neck cancers. More common in males (2:1). Tobacco and ethanol use are frequently causative factors.

Premalignant Lesions
Leukoplakia—white keratotic plaque that cannot be rubbed off, most common precursor to cancer

Erythroplakia—red mucosal plaque, greater risk for malignant transformation

Dysplasia—describes varying degrees of abnormal epithelial change, as dysplasia increases there is a greater malignant potential

Verrucous hyperplasia—irreversible premalignant lesion, histologically indistinguishable from verrucous carcinoma, associated with HPV

The "field cancerization" theory suggests that these patients are at increase risk for second primary lesion.

Pathology

Squamous cell carcinoma (SCC)—comprises > 90% of cancers in the oral cavity, ulcerative lesions are more common than exophytic

Basaloid squamous cell carcinoma—aggressive variant of SCC with increase recurrence and worse prognosis

Verrucous carcinoma—uncommon variant of SCC < 5%, warty tumor with pushing border

Adenoid cystic carcinoma— is the most common in the minor salivary gland

Predictors of Nodal Metastasis

Theses include tumor thickness (> 2 mm 40% incidence of cervical mets), degree differentiation and vascular or perineural invasion. The prognosis decreases with nodal disease and survival decreases by 50% with extracapsular spread in cervical metastasis.

Survival

Tongue 5 year survival—stage I 80%; II 68%; III 50%; IV 26%

Floor of mouth 5 year survival—stage I 80%; II 61%; III 28%; IV 6%

Cancer of the Oropharynx

This is less frequent, tends to be more poorly differentiated and is associated with tobacco, alcohol, prior radiation, and HPV. Patients present with symptoms of sore throat, otalgia, globus, odynophagia, dysphagia, trismus, weight loss, cervical lymphadenopathy. Treatment with surgery and radiation.

Cancer of the Upper Aerodigestive Tract

Presentation

Patients may present with pain, sore throat, otalgia, odynophagia, dysphagia, trismus, hoarseness, epistaxis, hearing change and/or weight loss.

Etiology

Tobacco (pack/year) use is a significant contributor to the development of these cancers as well as ethanol, immunodeficiency and Plummer-Vinson syndrome.

Exam

Complete exam including oral cavity, oral pharynx, hypopharynx, nasopharynx, and neck with direct visualization of area of concern with rigid nasal endoscopy or flexible laryngoscopy. Determine extent of tumor and function of vocal cords and must rule out second primary tumor.

Biopsy

A biopsy can be performed in the office using local if the tumor is accessible. If the tumor is in the nasopharynx a CT scan should be performed first before endoscopic biopsy is performed. Tumors in the hypopharynx or larynx require general anesthesia to evaluate the extent of the tumor and to take biopsies. Tracheostomy should be performed on any patient with an obstructing lesion. Open incisional biopsy of suspected neck metastasis should not be performed.

Imaging

CT scan with contrast of the neck from the skull base to the superior mediastinum is the best initial film and may show bony invasion or cervical metastasis. MRI can be used to evaluate tissue planes, airway lesions, or intracranial spread. Chest x-ray is mandatory for screeing for pulmonary metastasis.

Staging

Oropharynx T1 < 2 cm; T2 2-4 cm; T3 > 4 cm; T4 invades adjacent structure

Nasopharynx T1 one subsite; T2 >1 subsite; T3 invades nasal cavity; T4 invades skull

Hypopharynx T1 one site; T2 >1 site + nl larynx; T3 >1 site + fixed larynx; T4 invades adjacent structure

Supraglottis T1 1 site + nl larynx; T2 >1 site + nl larynx; T3 limited to larynx with fixed larynx or adjacent extension; T4 invasion

Glottis T1 limited to vocal cords, nl fxn; T2 extends above/below or abnl vocal cord fixation

T3 vocal cord fixation T4 invasion

Neck Nodal Staging

N1 < 3 cm

N2 3-6 cm

A ipsilateral to tumor, single

B ipsilateral to tumor, multiple

C contralateral

N3 > 6 cm

Salivary Gland Neoplasms

Background

The salivary glands are of ectodermal origin. As they develop they form acini, intercalated ducts, striated ducts and excretory ducts. Tumors arise from different aspects of the ducts to give the different tumor types. The risk factors for development of a salivary neoplasm include radiation therapy and Epstein-Barr virus infection.

Incidence

- 3-4% of head and neck tumors
- 54% of tumors are benign and 46% are malignant
- 75% of parotid neoplasms are benign

Salivary Gland Tumor Staging

T1= < 2 cm

T2= 2-4 cm

T3= 4-6 cm

T4= > 6 cm

Tumor-Like Conditions

Conditions that can mimic a neoplastic process are Sjogren's syndrome, brachial cleft cyst type I, vascular malformations, and HIV-associated parotid enlargement.

Parotid Gland

Anatomy

The parotid is the largest salivary gland, located between the ramus of mandible and mastoid tip. It has an asymmetric dumbbell shape with larger superficial lobe separated from a smaller deep lobe by the facial nerve. The gland is encapsulated by the parotid-masseteric fascia. The deep lobe is adjacent to the styloid process and the carotid sheath but can extend into the parapharyngeal space/lateral pharyngeal space. The facial nerve reliably originates from the stylomastoid foramen and is reliably 1 cm inferior and 1cm deep to the tragal cartilage. The parotid gland is innervated by the auriculotemporal nerve relay from CN IX. Stenson's duct drains parotid secretions into the oral cavity near upper 2nd molar.

Presentation

The most common presentation is a painless enlarging mass, pain or facial nerve weakness. This presentation is always worrisome for cancer.

Diagnosis

CT with contrast can be used but is not necessary for the diagnosis. MRI used to evaluate parapharyngeal spread or perineural invasion. FNA is usually not required in most tumors but does have good sensitivity (86%) and specificity (96-100%).

Benign Tumors

Pleomorphic Adenoma

It is derived from intercalated duct and myoepithelial cells. The majority occur in the parotid gland and is the most common tumor of the parotid gland. Approximately 96% occur in the superficial lobe and rarely transform into carcinoma. They are not well encapsulated. Pleomorphic adenomas contain pseudopod extensions which will recur if not adequately excised.

Warthins Tumor

These may occur during embryogenesis due to incorporation of salivary ducts and lymphatic tissue. Warthin's tumor is also known as adenolymphoma due to its lymph node architecture and epithelial components. It is the second most common benign tumor of the parotid (10-15%). Approximately 10% are bilateral. They are more common in males with a weak relation to smoking.

Other Benign Tumors Include

- monomorphic adenoma
- basal cell adenoma
- oncocytoma
- clear cell adenoma
- hemangioma in children

Carcinoma of the Parotid

Mucoepidermoid Carcinoma

It is the most common malignant tumor of parotid and salivary glands responsible for 35% of salivary tumors. They are of mucin and epithelial cell derivation.

They are subdivided into:
- Low grade—well differentiated
- High grade—poorly differentiated

High grade has a high incidence of locoregional recurrence, distant metastasis, and shorter survival. Over 70% have lymphatic metastasis. High grade—treat with superficial/total parotidectomy with facial nerve sacrifice if necessary and if no clinical nodal disease do a selective nodal dissection. If palpable nodal disease is present, a formal neck dissection is performed.

Low grade—treat with superficial lobectomy and spare facial nerve unless directly involved; then resect and cable graft it with donor nerve.

Prognosis
5 year survival 70% low grade and 47% high grade

Acinic Cell Carcinoma

It accounts for 13-16% of parotid malignancies. Approximately 81% occur in the parotid, 4% in submandibular gland, 13% in minor salivary glands. This is considered a low grade malignancy. They are often solitary, encapsulated with well-defined margins.

Histologic Subtypes
- solid
- microcystic
- papillary-cystic
- follicular

The histologic differentiation doesn't predict survival.

Management
Treatment with wide local resection with facial nerve preservation if possible.

Prognosis
10 year survival 85%

Adenoid Cystic Carcinoma

Represents 10% of all salivary neoplasms. The oral cavity is the most common site (50%), sinonasal (18%), submandibular/parotid (28%),and lingual (1%).

Histologic Subtypes
- cribiform (44%)
- tubular
- solid (21%)

Cribiform has the best prognosis (100% 5 year survival). Solid has worst prognosis, with a 17% 5 year survival. Perineural invasion occurs in 20-80% of tumors.

Management

Treatment with wide local excision, check nerve margins for invasion and can treat with adjuvant radiation therapy.

Malignant Mixed Tumor

Most common in the parotid representing 3-12% of salivary gland cancer. Three to four percent of pleomorphic adenomas progress to this carcinoma with a risk of transformation 10% after 15 years. Treat with wide local resection with facial nerve preservation if possible. Overall there is a poor survival with 5 year 40%, 10 year 24%, 15 year 19%.

Submandibular Gland

Fills the submandibular triangle and Wharton's duct drains saliva into the floor of the mouth. Its blood supply is from the facial artery. The majority of masses here are malignant 86% and present with swelling and possibly pain. Treat with excision via an incision two fingerbreadths below the mandible to avoid injury to the facial nerve

Minor Salivary Gland Tumors

Usually presents as a smooth submucosal mass located in submucosa throughout oral cavity. They produce mucinous secretions, and the majority of tumors are malignant. The majority of malignant tumors are adenoid cystic carcinomas. Treatment is wide local excision depending on the site.

CHAPTER 23

Pediatric Surgery
Monica S. Hall

Pediatric Surgery Overview

Management of Fluids and Electrolytes

Maintenance IV fluids can be written using one of two formulas:

The 4-2-1 Rule is written in mL/hr and is as follows; 4 mL/kg/hr up to 10 kg of body weight, additional 2 mL/kg/hr for each kg above 10 and up to 20 kg, additional 1 mL/kg/hr for each kg above 20 kg body weight.

The Daily Maintenance Fluid Formula for total maintenance IV fluids in a day (which should be divided by 24 to give an hourly rate) is as follows; 100 mL/kg up to 10 kg of body weight, additional 50 mL/kg for each kg above 10 and up to 20 kg, additional 25 mL/kg for each kg above 20.

Daily requirements are 5 mEq/kg/day of sodium, 2 mEq/kg/day of potassium, and 2 mEq/kg/day of calcium. Blood product replacement in pediatrics is by weight. Infant blood volume is 85 mL/kg of body weight

- PRBC 10 mL/kg
- FFP 20 mL/kg
- Platelets 1U/5kg

Lesions of the Neck

Branchial Cleft Anomalies

The branchial clefts and arches develop during the fourth week of gestation. Remnants of these structures result in branchial cleft cysts and sinuses. The most common anomalies arise from the second branchial cleft.

Second branchial cleft sinus is found as an opening from the anterior border of the lower one-third of the sternocleidomastoid muscle. It is most commonly associated with clear drainage, and its presentation is usually in childhood. Second branchial cleft cysts usually present as a mass that is anterior to and below the upper one third of the SCM. It presents in adulthood and there is an associated risk of in situ cancer.

Treatment of these anomalies consists of removal of the entire cyst and/or tract. In the setting of infection, initial treatment should consist of adequate incision and drainage (10% of cases are bilateral).

Thyroglossal Duct Remnants

Thyroglossal duct cysts result from thyroid tissue which fails to completely migrate during embryologic descent of the thyroid from its original position at the foramen cecum / base of tongue. It usually presents in preschool aged children as an asymptomatic midline mass at or below the level of the hyoid bone. It is found to

Current Concepts in General Surgery: A Resident Review, edited by William R. Wrightson.
©2006 Landes Bioscience.

move when the tongue is protruded. Occasionally, symptomatic lesions may present as pain, or difficulty with swallowing.

Treatment of a thyroglossal duct cyst consists of the Sistrunk operation, which is removal of the cyst to include the central portion of the hyoid bone and the tract to the foramen cecum, where it is ligated. In the setting of infection, acute treatment should consist of adequate drainage and intravenous antibiotics. There is a risk of malignant degeneration of thyroid tissue contained within the cyst if the remnant is left until adulthood.

Cervical Adenitis

Cervical adenitis is associated with pharyngitis or otitis media caused by staphylococcus or streptococcus species, and therefore treatment is directed at the primary cause (antibiotics for the original infection). Occasionally, fluctuance of lymph nodes may require incision and drainage. Chronic cervical adenitis may be associated with cat scratch fever, atypical mycobacterium, or tuberculosis.

Cystic Hygroma

Cystic hygroma is caused by obstruction of the lymphatic vessels during their development, which results in endothelium lined cysts which are filled with lymph. This condition usually presents by the age of two years, and the most commonly affected area is the posterior triangle of the neck. It can also form in the groin, axilla, and mediastinum. Treatment involves excision, which may be limited by its involvement with surrounding structures.

Chest Lesions

Bronchogenic Cysts

Bronchogenic cysts are hamartomatous cysts which are formed from abnormal embryologic budding of the tracheobronchial tree. The result is a cyst with a lining which contains respiratory epithelium, smooth muscle, and cartilage. These cysts usually present as asymptomatic lesions in any location in a child of any age. The location of the cysts determines the symptoms, if any. A cyst can give rise to compressive symptoms if it is within the trachea. Treatment involves surgical removal of the cyst.

Congenital Lobar Emphysema (CLE)

An obstructive process which results in progressive expansion of the lobe(s) distal to it. The hyperinflated segment then compresses the surrounding lung tissue. It is most commonly caused by inadequate bronchial cartilage to provide support, but may also result from extrinsic compression. The most commonly affected area is the left upper lobe, then right upper lobe, then right middle lobe. The usual presentation is in a newborn as dyspnea, wheezing, and cough. Workup should not include bronchoscopy as it can cause further air trapping and worsening symptoms. Treatment usually involves lobectomy.

Congenital Cystic Adenomatoid Malformation

Cystic and/or solid proliferation of immature lung tissue at the terminal airways can result in air trapping with compression of the surrounding lung parenchyma, much like CLE. The most commonly affected area is the left upper lobe of the lung. It often presents at birth as respiratory difficulty, and treatment is resection, usually a lobectomy.

23

Pulmonary Sequestration

Pulmonary sequestration involves a portion of lung tissue which is not in communication with the pulmonary arterial or bronchial tree.

Intralobar sequestrations are within the lung tissue, usually of the left lower lobe, and are diagnosed in childhood. These lesions may have a secondary connection to the tracheobronchial tree. Treatment is usually lobectomy and is reserved for patients who are symptomatic.

Exralobar sequestrations are outside of the main lung parenchyma, usually posterior and left, and usually have their own visceral pleura. These lesions are typically diagnosed in infancy and treatment is simple excision if symptomatic.

Congenital Diaphragmatic Hernia

Caused by failed closure of the pleuroperitoneal canal, which is the last portion of the diaphragm to close. It therefore usually occurs on the left side, posterior and laterally. Presentation is as respiratory distress at birth or several hours thereafter and is due to pulmonary hypoplasia, pulmonary artery hypertension, and/or persistence of fetal circulation. Radiographically it may be seen as air-filled bowel within the chest cavity. Initial treatment is stabilization and ventilation/oxygenation by ventilator support or by ECMO. Surgical repair involves a two layered primary closure of the defect. Occasionally, a prosthetic patch may be required.

Foreign Bodies

Airway foreign bodies usually lodge in the right mainstem bronchus or the RLL. Bronchoscopy is required for removal of the foreign body. Esophageal foreign bodies usually lodge at one of three points of narrowing: the cricopharyngeus, the aortic arch, or the GE junction. Esophagoscopy is required for removal.

Esophageal Atresia and Tracheoesophageal Fistulas

If the trachea and esophagus fail to completely separate by 35 days of gestation, it can result in one of several anomalies. The most commonly seen type (85%) is that of a blind upper esophageal pouch and a distal esophageal segment which is connected to the trachea (Type C). It usually presents as abdominal distension, feeding intolerance, and pneumonia in the newborn. As many as half of these cases may be associated with other components of the VATER complex of anomalies (vertebral, anal, tracheoesophageal, and radial/renal). Treatment involves closure of the fistula with primary anastomosis of the esophagus without tension. In the unstable infant, repair is delayed and initial treatment is G tube placement with division of the TE fistula.

Corrosive Esophageal Injuries

Upper endoscopy is used to further evaluate these injuries, but the endoscope should never be advanced past the first level of the burned area due to the elevated risk of perforation. Antibiotics are given for three weeks. Subsequent stricture formation may require esophageal dilatation, which should not be done less than three weeks from the time of the injury.

Gastroesophageal Reflux

Gastroesophageal reflux should be suspected if there is frequent emesis and obstruction has been ruled out. Its workup involves barium swallow and 24 hour pH probe. Initial treatment should involve thickening the formula with rice cereal and keeping the patient upright while eating. Medicinal management includes H2 blockers and prokinetic agents. Surgical therapy is reserved for those in which conservative measures prove inadequate or who have complications such as stricture formation or bleeding. Surgery involves the Nissen or Thal fundoplication.

Lesions of the Abdomen/GI Tract

23

Pyloric Stenosis

Pyloric stenosis typically presents in first born males, aged 4-8 weeks, as increasingly projectile nonbilious emesis. It is caused by muscular hypertrophy of the pylorus, which results in outflow obstruction. Physical exam may reveal a classic palpable olive in the right upper quadrant or upper mid abdomen. Electrolyte abnormalities are common and include metabolic alkalosis with potassium and chloride depletion. Diagnosis can be confirmed with ultrasound or upper GI series. Treatment involves fluid resuscitation and correction of electrolyte abnormalities, followed by the Fredet-Ramstedt pyloromyotomy, which is incision and spreading open of the seromuscular layer of the pylorus. Postoperative feeding protocols involve frequent small feeds of progressively less dilute formula over approximately two days.

Duodenal Obstruction

Duodenal obstruction results from duodenal atresia, web, stenosis, annular pancreas, or malrotation. It usually presents in newborns as feeding intolerance and stomach distension. Bilious emesis is seen if the obstruction is distal to the ampulla of Vater, which is the case in the vast majority of cases. One-third of patients will also have Down's syndrome. Diagnosis is confirmed with air contrast upper GI with findings of a "Double-Bubble". Initial treatment is NG tube decompression and IVF hydration. Operative treatment is dependent upon the cause of the obstruction but should be pursued rapidly if malrotation is the suspected cause. The most commonly employed surgical therapy is duodenoduodenostomy. However, duodenal webs may be treated with duodenotomy and excision. An annular pancreas should never be divided.

Intestinal Atresia

Intestinal atresia is commonly caused by fetal mesenteric vascular accidents and presents at birth as bilious emesis, progressive abdominal distension, and failure to pass meconium. Workup includes contrast enema as well as suction rectal biopsy to exclude Hirschsprung's disease. There are four types of intestinal atresia. Type I is mucosal atresia only. Type II involves atretic ends connected by a fibrous band of tissue. Type IIIA, the most common type, has two atretic ends of bowel separated by a V shaped defect in the mesentery. In Type IIIB, the distal atretic bowel receives blood from the ileocolic or right colic arteries, also referred to as the "apple peel" type. Type IV involves multiple atretic segments. Surgical therapy of intestinal atresia should be pursued urgently and involves resection of the affected bowel with primary anastomosis.

Malrotation and Midgut Volvulus

During normal development, the midgut undergoes a 270° counterclockwise rotation along the axis of the superior mesenteric artery. When this rotation is incomplete, problems can arise because of duodenal obstruction from Ladd bands or midgut volvulus. Their most common clinical presentation is that of bilious emesis in the newborn, but they occasionally may take up to one year to become apparent clinically. Diagnosis is confirmed by UGI study to assess if the duodenum crosses the vertebral column. Surgical therapy is pursued emergently and involves detorsion of the midgut in a counterclockwise direction, division of Ladd bands, passage of a rubber catheter to rule out duodenal obstruction, and replacement of the cecum in the LUQ and the duodenum in the RUQ, and appendectomy. If laparotomy reveals ischemic bowel, a second laparotomy is performed after 24 to 36 hrs to assess the viability of the bowel.

Meconium Ileus

Obstruction can be caused by meconium impaction within the ileum. It is most commonly encountered in newborns with cystic fibrosis. Workup involves a contrast enema which may reveal a microcolon and meconium within the distal ileum. If there are no complicating problems, it is appropriate to attempt conservative therapy first, which includes gastrografin enemas until the meconium passes. Surgical therapy is instituted when conservative measures fail, or there are complications such as perforation. It includes resection of the distended terminal ileum with ostomy or end-to-end anastomosis.

Intussusception

Intussusception is caused most commonly by hypertrophied Peyer's patches in the terminal ileum, followed by polyps, lymphoma or other tumors, and Meckel's diverticulum. It commonly presents in infants between 8 months to 1 year of age as intermittent paroxysms of crampy abdominal pain and emesis. The patient may feel well between these episodes. Diagnosis is confirmed by x-ray which reveals a RUQ or midepigastric mass with an "empty" RLQ (Dance's Sign). Treatment is dependent on the progression of symptoms. If no peritoneal signs are present, air enema is performed in an attempt to reduce the intussusception. Post-reduction x-ray should reveal reflux of air into multiple small bowel loops and resolution of symptoms. Failure of conservative therapy mandates surgery which involves gentle milking of the bowel out of the intussusception. If the affected bowel is gangrenous, resection is performed.

Necrotizing Enterocolitis (NEC)

Necrotizing enterocolitis is most commonly seen in premature infants and presents as intolerance of feeds which progresses to distension and bloody stools. It may be associated with ischemia, malnutrition, and use of synthetic formulas. Radiographic findings may include pneumatosis intestinalis which is caused by gas forming bacteria. Initial treatment is to stop the tube feeds, place an NG for decompression, and start TPN and antibiotic therapy. Surgery involves resection of the affected bowel with ostomy and is performed on those infants with free air, refractory acidosis, peritoneal signs, or gas within the portal vein among other findings. If laparotomy reveals massive gangrene/necrosis, only the bowel which is clearly necrotic is resected, and a reexploration is performed 24 hours later. Another option is that of intraabdominal drain placement, which is often reserved for infants weighing less than 1.5 kg. Survival is approximately 80% overall.

Appendicitis

It's presentation may be confused with gastroenteritis but usually includes one or all of the following; anorexia, fever, umbilical pain which migrates to the right lower quadrant. Most patients present within one to two days of the onset of symptoms, though many exceptions exist. Diagnosis may be confirmed with abdominal CT scan or contrast enema and are usually reserved for cases which are indeterminate.

23

Intestinal Duplications

These are cystic masses which often have a common wall with the small intestine, usually the ileum. May be asymptomatic, or may present as an abdominal mass, or as bleeding or obstruction. Diagnosis is confirmed with technetium pertechnetate scan. Type of surgical treatment pursued depends upon the length of intestine which is involved with the cyst. If it is short, resection of the cyst and its associated bowel with reanastomosis is performed. If the involved segment of bowel is long, the mucosa of the duplication should be stripped, allowing the cyst to collapse, or stapling of the common wall may be effective.

Meckel's Diverticulum

This is usually found on the antimesenteric side of the ileum and is a remnant of the vitelline duct. It is the most common cause of lower GI bleeding in children. Presentation ranges from asymptomatic to pain and bleeding. There is ectopic gastric mucosa in approximately half of all Meckel's. The Rule of Twos states that it is usually two inches in length, within two feet of the ileocecal valve, two percent of patients are symptomatic, and two times more common in males. It's diagnosis can be confirmed by technetium pertechnetate scan. Surgical therapy involves wedge resection with transverse closure or sleeve excision and end-to-end anastomosis if the base of the diverticulum is wide.

Mesenteric Cysts

Mesenteric cysts are most commonly located in the region of the ileum and do not contain mucosa or muscle. They may present as an obstruction, or as an abdominal mass. Diagnosis is by ultrasound or abdominal CT scan, and treatment involves resection which may require a bowel resection.

Hirschsprung's Disease

Hirschsprung's disease is a condition caused by the absence of ganglion cells in the myenteric plexus which causes a mechanical obstruction. The most commonly affected area is the rectosigmoid region. Its typical presentation is as a failure to pass meconium during the first 24 hrs after birth, or it can also present as chronic constipation during the first year after birth. Diagnosis is confirmed by suction rectal biopsy with pathologic findings of the absence of ganglion cells in the myenteric plexus, increased cholinesterase staining and nerve bundle hypertrophy. Treatment involves resection of the affected bowel, with frozen section biopsy of the ostomy to ensure that there are ganglion cells present. Pull-through anastomosis is performed approximately 3 to 6 months later.

Imperforate Anus

Imperforate anus is caused by failure of the descent of the urorectal septum. High versus low imperforate anus is determined by where the rectum ends, above or below the levator ani. There is a high incidence of associated genitourinary anomalies, especially with high imperforate anus (VATER, VACCTERL). Diagnosis of this condition is by careful physical exam of the perineum, forchette, and vestibule. Low imperforate anus will present as a fistula that ends at the posterior forchette in females and at the median raphe of the penis or scrotum in males. Radiographic evaluation includes lateral x-ray with the buttocks up or pelvic x-ray with the infant held upside down. Treatment of high imperforate anus is colostomy with later posterior sagittal anorectoplasty, and for low imperforate anus, treatment may be performed by a perineal approach.

Biliary Atresia

An obliterative process which involves the common duct, cystic duct, one or both hepatic ducts and the gallbladder. It results in pathologic jaundice (newborn jaundice lasting greater than 2 weeks, with a direct BR of greater than 2 mg/dl). Polysplenia is a commonly associated anomaly, among others. Evaluation involves fasting abdominal ultrasound, needle biopsy of the liver, checking the alpha-1 antitrypsin levels, and technetium 99mIDA scan after pretreatment with phenobarbital. Surgical treatment is the Kasai hepatoportoenterostomy, which is based on the observation that fibrous tissue of the porta hepatis has microscopically patent biliary ducts that communicate with the intrahepatic ductal system. This procedure includes transection of these patent ducts and placing a Roux en y limb of jejunum in communication with them. This can be complicated by postoperative cholangitis. Cure rate with the Kasai procedure is approximately 25-30%. Liver transplantation is performed if the Kasai procedure fails. Biliary atresia is the most common indication for liver transplantation in the pediatric population.

Umbilical Hernia

A failed closure of the umbilical ring causes a defect in the abdominal wall. For those with a fascial defect of less than 1 cm, closure will usually occur by the age of 3, and it is prudent to wait until the age of 5 if the defect is small enough to close spontaneously. Surgical treatment is simple repair with interrupted sutures in a transverse plane. Incarceration is rare with umbilical hernias.

Inguinal Hernia

Pediatric inguinal hernias usually result from failed closure of the processus vaginalis which is most commonly seen in males. Most of these indirect hernias require only high ligation of the hernia sac, and have a remarkably low recurrence rate.

Omphalocele

Probably secondary to an interruption of central migration of the lateral folds during abdominal wall development. There is a high incidence of associated anomalies, especially cardiac. Treatment initially is by moist covering. A temporary silo of silastic can be used to provide steady reduction when the abdomen cannot be closed primarily. It should be reduced over the course of approximately 7 days, and excess tension should be avoided.

Gastroschisis

Gastroschisis involves a full-thickness defect of the abdominal wall which is believed to be formed secondary to intrauterine rupture of a hernia of the umbilical cord. It is almost always found to the right of the umbilicus, and there is no associated peritoneal sac. It can also be associated with bowel atresia. Urgent surgical therapy involves primary closure when possible or silastic covering with gradual reduction, being mindful to avoid excess intraabdominal pressure.

Prune-Belly Syndrome (Eagle-Barret)

Prune Belly syndrome is a laxity of the lower abdominal musculature associated with dilation of the urinary system, and undescended testicles, most commonly affecting males only.

23

Neoplasms

Wilms Tumor

An embryonal neoplasm of the kidney which usually presents in children between the ages of 1 and 5 years as a mass of the upper abdomen or flank. It may be associated with such anomalies as aniridia, Beckwith-Wiedemann syndrome, urinary tract defects, hemihypertrophy, and chromosomal deletion. Its workup includes CT of the abdomen and chest and ultrasound which are important to rule out involvement of the renal vein or IVC. Surgical treatment is the primary mode of therapy and involves clamping the renal vein first to prevent venous embolization of tumor cells. Lymph nodes are sampled at the time of resection to assist with staging, which will determine if chemotherapy or radiation therapy will be needed. Patients with disease limited to the kidney which is completely excised, and treated with chemotherapy.

Neuroblastoma

Cancer which arises from neural crest cells, which most commonly occur in preschool aged children. Tumor location may be the adrenal glands, neck, pelvis, mediastinum, or any other place where sympathetic ganglion cells are located. It usually presents as an asymptomatic abdominal mass. Workup includes CT and ultrasound, followed by possible BM biopsy and radionuclide scanning. Laboratory evaluation includes blood and urine levels of VMA and HVMA. The most effective treatment is surgical excision.

Rhabdomyosarcoma

A soft tissue tumor which can arise anywhere, most commonly in the head and neck. Evaluation is by biopsy, MRI, CT and BM biopsy as the tumor spreads by both local extension and distantly. Treatment is wide local excision, as well as chemotherapy, and possible radiation therapy, and is dependent upon the location and histology of the tumor.

Teratoma

Tumors containing all three types of embryonal germ tissue. Usually, they are found in the midline and treatment involves complete surgical excision.

Sacrococcygeal teratomas are most often seen as sacral masses in the newborn. For those with a large sacrococcygeal tumor and hydrops, the outlook is poor. There is a potential for malignant degeneration as the child grows, and therefore complete excision of the tumor is indicated.

Liver Masses

Most liver tumors in children are malignant, with hepatoblastoma being the most common. Hepatocellular carcinoma is the next most common and has a poorer prognosis. The most common presenting complaint is that of weight loss and an abdominal mass. Workup includes plain x-rays as well as ultrasound and contrasted CT scan. Treatment is by surgical excision when possible. Cure rates for hepatoblastoma treated by resection with chemotherapy are much greater than those for hepatocellular carcinoma.

References

1. Guzzetta PC, Anderson KD, Altman RP et al. Principles of surgery. 7th ed. USA: McGraw-Hill, 1999:1715-1754.
2. Kosloske AM. Surgery of Infants and Children: Scientific principles and practice. Philadelphia: Lippincott-Raven, 1997:1201-1212.
3. Bond SJ, Groff DB. Pediatric surgery. 5th ed. St. Louis: Mosby, 1998:2:1257-1267.
4. Reynolds M. Swenson's pediatric surgery. 5th ed. New York: Appleton and Lange, 1990:721-735.

I

J

Index

Index